A General Algebraic Semantics for Sentential Logics
Second Edition

Since their inception, the Perspectives in Logic and Lecture Notes in Logic series have published seminal works by leading logicians. Many of the original books in the series have been unavailable for years, but they are now in print once again.

In this volume, the 7th publication in the Lecture Notes in Logic series, Font and Jansana develop a very general approach to the algebraization of sentential logics and present their results on a number of particular logics. The authors compare their approach, which uses abstract logics, to the classical approach based on logical matrices and the equational consequence developed by Blok, Czelakowski, Pigozzi and others. This monograph presents a systematized account of some of the work on the algebraic study of sentential logics carried out by the logic group in Barcelona in the 1970s.

JOSEP MARIA FONT works in the Department of Probability, Logic and Statistics at the University of Barcelona.

RAMON JANSANA works in the Department of Logic, History and Philosophy of Science at the University of Barcelona.

LECTURE NOTES IN LOGIC

A Publication of The Association for Symbolic Logic

This series serves researchers, teachers, and students in the field of symbolic logic, broadly interpreted. The aim of the series is to bring publications to the logic community with the least possible delay and to provide rapid dissemination of the latest research. Scientific quality is the overriding criterion by which submissions are evaluated.

More information, including a list of the books in the series, can be found at http://www.aslonline.org/books-lnl.html

LECTURE NOTES IN LOGIC 7

A General Algebraic Semantics for Sentential Logics

Second Edition

JOSEP MARIA FONT

University of Barcelona

RAMON JANSANA

University of Barcelona

ASSOCIATION FOR SYMBOLIC LOGIC

CAMBRIDGE
UNIVERSITY PRESS

CAMBRIDGE
UNIVERSITY PRESS

University Printing House, Cambridge CB2 8BS, United Kingdom

One Liberty Plaza, 20th Floor, New York, NY 10006, USA

477 Williamstown Road, Port Melbourne, VIC 3207, Australia

4843/24, 2nd Floor, Ansari Road, Daryaganj, Delhi – 110002, India

79 Anson Road, #06–04/06, Singapore 079906

Cambridge University Press is part of the University of Cambridge.

It furthers the University's mission by disseminating knowledge in the pursuit of education, learning, and research at the highest international levels of excellence.

www.cambridge.org
Information on this title: www.cambridge.org/9781107167971
10.1017/9781316716915

First edition © 1996 Springer-Verlag Berlin Heidelberg
Second edition © 2009 Association for Symbolic Logic
This edition © 2016 Association for Symbolic Logic under license to
Cambridge University Press.

Association for Symbolic Logic
Richard A. Shore, Publisher
Department of Mathematics, Cornell University, Ithaca, NY 14853
http://www.aslonline.org

A catalogue record for this publication is available from the British Library.

ISBN 978-1-107-16797-1 Hardback

Cambridge University Press has no responsibility for the persistence or accuracy of URLs for external or third-party Internet Web sites referred to in this publication and does not guarantee that any content on such Web sites is, or will remain, accurate or appropriate.

CONTENTS

INTRODUCTION 1

CHAPTER 1. GENERALITIES ON ABSTRACT LOGICS
            AND SENTENTIAL LOGICS 15

CHAPTER 2. ABSTRACT LOGICS AS MODELS
            OF SENTENTIAL LOGICS 31
    2.1.   Models and full models 31
    2.2. S-algebras 36
    2.3.   The lattice of full models over an algebra 40
    2.4.   Full models and metalogical properties 45

CHAPTER 3. APPLICATIONS TO PROTOALGEBRAIC
            AND ALGEBRAIZABLE LOGICS 59

CHAPTER 4. ABSTRACT LOGICS AS MODELS
            OF GENTZEN SYSTEMS 75
    4.1.   Gentzen systems and their models 76
    4.2.   Selfextensional logics with Conjunction 86
    4.3.   Selfextensional logics having the Deduction Theorem 95

CHAPTER 5. APPLICATIONS TO PARTICULAR
            SENTENTIAL LOGICS 105
    5.1.   Some non-protoalgebraic logics 107
    5.1.1. CPC$_{\wedge\vee}$, the $\{\wedge,\vee\}$-fragment of Classical Logic 107
    5.1.2. The logic of lattices 110
    5.1.3. Belnap's four-valued logic, and other related logics 111
    5.1.4. The implication-less fragment of IPC and its extensions 113
    5.2.   Some Fregean algebraizable logics 114
    5.2.1. Alternative Gentzen systems adequate for IPC$_{\rightarrow}$ not having the
            full Deduction Theorem 116

CONTENTS

5.3. Some modal logics 117
5.3.1. A logic without a strongly adequate Gentzen system 121
5.4. Other miscellaneous examples 121
5.4.1. Two relevance logics 122
5.4.2. Sette's paraconsistent logic 123
5.4.3. Tetravalent modal logic 125
5.4.4. Cardinality restrictions in the Deduction Theorem 126

BIBLIOGRAPHY 131

SYMBOL INDEX 143

GENERAL INDEX 147

INTRODUCTION

The purpose of this monograph is to develop a very general approach to the algebraization of sentential logics, to show its results on a number of particular logics, and to relate it to other existing approaches, namely to those based on logical matrices and the equational consequence developed by Blok, Czelakowski, Pigozzi and others.

The main distinctive feature of our approach lies in the mathematical objects used as models of a sentential logic: We use *abstract logics*[1], while the classical approaches use *logical matrices*. Using models with more structure allows us to reflect in them the metalogical properties of the sentential logic. Since an abstract logic can be viewed as a "bundle" or family of matrices, one might think that the new models are essentially equivalent to the old ones; but we believe, after an overall appreciation of the work done in this area, that it is precisely the treatment of an abstract logic as a single object what gives rise to a useful—and beautiful—mathematical theory, able to explain the connections, not only at the logical level but at the metalogical level, between a sentential logic and the particular class of models we associate with it, namely the class of its *full models*.

Traditionally logical matrices have been regarded as the most suitable notion of model in the algebraic studies of sentential logics; and indeed this notion gives several completeness theorems and has generated an interesting mathematical theory. However, it was not clear how to use the matrices in order to associate a class of algebras with an arbitrary sentential logic, in a general way that could be mathematically exploited in order to find and study the connections between the properties of the sentential logic and the properties of the class of algebras; and this was true in spite of the fact that in most of the best-known logics these connections were recognized early. Rasiowa singled out in her [1974] the *standard systems of implicative extensional propositional calculi*, based on an implication

---

[1] In our own later publications we have preferred the term *generalized matrices* over that of *abstract logics*, in order to avoid any misunderstsanding with concepts in abstract model theory. See Font [2003b] and Font, Jansana, and Pigozzi [2001], [2003], [2006].

connective, and Czelakowski studied in his [1981] the much more general *equivalential logics*, based on the behaviour of a generalized equivalence connective.

In the late eighties two fundamental papers by Blok and Pigozzi decisively clarified some points; in their [1986] they introduced *protoalgebraic logics*, and in their [1989a] they introduced a very general notion of what an "algebraic semantics" means, and defined the *algebraizable logics*. With each algebraizable logic there is associated a class of algebras, its *equivalent quasivariety semantics*, in such a close way that the properties of the consequence relation of the logic can be studied by looking at the properties of the equational consequence relative to the class of algebras and vice-versa; the links between logic and algebra, expressed by means of two elementary definable translations, are here very strong. The paradigmatic examples of algebraizable logics are classical and intuitionistic propositional calculi, whose equivalent quasivariety semantics are Boolean and Heyting algebras respectively. Protoalgebraic logics form a wider class of sentential logics, and they also have an associated class of algebras, the *algebra reducts of their reduced matrices*, but for these logics it is not the class of algebras but *the class of matrices* what has a good behaviour in its relationship with the logic; that is, its behaviour is somehow analogous to that of the equivalent quasivariety semantics for algebraizable logics, and many of the relevant theorems of universal algebra have an analogue for matrices of protoalgebraic logics. One paradigmatic example of a protoalgebraic but non-algebraizable logic is the sentential logic obtained from the normal modal logic S5 by taking all its theorems as axioms and Modus Ponens as the only rule of inference from premisses. Up to now, protoalgebraic logics seem to form the widest class of sentential logics which are "amenable to most of the standard methods of algebraic logic" (Blok and Pigozzi [1989a] p. 4). And only for algebraizable logics does the common phrase "these algebras play for this logic a similar role to that played by Boolean algebras for classical logic" make real and full sense.

However, algebraizable and protoalgebraic logics are not the only ones of interest; others[2] are the $\{\wedge, \vee\}$-fragment of classical logic, studied in Font and Verdú [1991]; the implication-less fragment of intuitionistic propositional logic, studied in Rebagliato and Verdú [1993]; and Belnap's four-valued logic, studied in Font [1997] (they are also dealt with, respectively, in Sections 5.1.1, 5.1.4 and 5.1.3 of the present monograph). These logics are associated in a natural way with

---

[2]After 1996 a few other logics have been indentified as non-protoalgebraic: Certain subintuitionistic logics treated in Bou [2001] and in Celani and Jansana [2001]; some positive modal logics studied in Jansana [2002]; and a large family of logics that preserve degrees of truth related to many-valued logic and to varieties of residuated structures, studied in Font [2003a], Font, Gil, Torrens, and Verdú [2006] and Bou, Esteva, Font, Gil, Godo, Torrens, and Verdú [2009].

a class of algebras (the distributive lattices, the pseudo-complemented distributive lattices, and the De Morgan lattices, respectively); but it turns out that these are not the classes of algebras that the traditional matrix approach would associate with them, that is, they are not the algebra reducts of their reduced matrices, as proved in Font, Guzmán, and Verdú [1991], in Rebagliato and Verdú [1993] and in Font [1997], respectively. However, these classes of algebras can be characterized by the structure of the set of their deductive filters, namely by the fact that the abstract logic associated with this set satisfies some typical metalogical properties, also characteristic of the corresponding logic. So we find that, if instead of matrices we use abstract logics with some special properties as the models of the logics, then we can characterize the associated algebras as the algebra reducts of the reduced models.

The procedure just described can be generalized. We associate with each sentential logic S a class of abstract logics called *the full models of S* (Definition 2.8) with the conviction that (some of) the interesting metalogical properties of the sentential logic are precisely those shared by its full models. With the aid of the full models we associate with any sentential logic S a class of algebras, called the class of *S-algebras*, which are the algebra reducts of the reduced full models. And we claim that *the notion of full model is a "right" notion of model of a sentential logic*, and, even more emphatically, that *the class of S-algebras is the "right" class of algebras to be canonically associated with a sentential logic*. To support these claims we offer three groups of reasons: In the first place, there are the general results we prove in the monograph, especially in Chapter 2, which seem of interest by themselves, but also due to their applications in the theory of protoalgebraic and algebraizable logics, as the contents of Chapters 3 and 4 show. Second, the application of our general constructions to the study of many particular logics, which are dealt with in Chapter 5; we have examined a variety of sentential logics and found that the class of S-algebras is always the "right" one, i.e., the one expected by other, sometimes partial or unexplained connections. And third, the fact that our proposal is consistent with previous ones, since in all cases where an alternative approach exists, the class of algebras it associates with a sentential logic is also the class of S-algebras: this is so for the protoalgebraic and the algebraizable cases (see Proposition 3.2), and also for many sentential logics defined by a Gentzen system which is "algebraizable" in the sense of Rebagliato and Verdú [1993], [1995]. In Chapter 4 we see that this consistency also extends to the associated abstract logics: Under reasonable restrictions on S, the classes of abstract logics and of algebras found by using the notion of model of a Gentzen system are also the full models of S and the S-algebras, respectively;

and moreover, for a class of sentential logics which includes all the algebraizable ones, the matrices and the full models can essentially be identified by the
isomorphism exhibited in Theorem 3.8, a completely natural one.

This monograph can also partly be seen as an attempt to present a systematized
account of some of the work on the algebraic study of sentential logics using
abstract logics carried out by several people in Barcelona since the mid-seventies.
It is not a retrospective survey (the Barcelona group has produced other work
following different lines of research in the field of Algebraic Logic) but rather an
attempt to build a general framework that both explains and generalizes many of
the results obtained in this area, and makes it possible to connect them with other
(older or newer) approaches to the algebraization of logic. Thus, the contents of
this monograph cannot be properly motivated without these references; since our
approach is not yet standard, it may be interesting, or even necessary, to detail
some elements of its historical development; see also Font [1993], [2003b].

Some history

Abstract logics are pairs $\langle A, \mathrm{C} \rangle$ where A is an algebra and C is a closure
operator defined on the power set of its universe. Dually, they can be presented as
pairs $\langle A, \mathcal{C} \rangle$ where $\mathcal{C}$ is the closure system associated with the closure operator
C (see page 17); as such they have been called *generalized matrices* by Wójcicki,
who in Section IV.4 of his [1973] points out that each one of them is equivalent,
from the semantical standpoint, to a family of logical matrices, and that "[this
notion] does not provide us with essentially new tools for semantical analysis
of sentential calculi". However, the notion of closure operator incorporates a
qualitatively different element of logic, namely, the possibility of expressing, in
abstract form, some metalogical properties of the operation of logical inference;
the best known of these is the Deduction Theorem: $\Gamma, \varphi \vdash_S \psi \iff \Gamma \vdash_S \varphi \to \psi$,
which can be written as $\psi \in \mathrm{Cn}_S(\Gamma \cup \{\varphi\}) \iff \varphi \to \psi \in \mathrm{Cn}_S(\Gamma)$, where Cn_S
is the closure operator corresponding to the consequence relation $\vdash_S$ associated
with the logic S (that is, $\varphi \in \mathrm{Cn}_S(\Gamma) \iff \Gamma \vdash_S \varphi$).

We believe that it is fair to say that the study of the properties of the closure
operators (also called *consequence operators* in this context) of logical systems
starts with Tarski [1930], where he even *defines* classical logic as (in today's
words) a closure operator on the algebra of sentential formulas satisfying some
metalogical properties like being finitary, the Deduction Theorem for implication, and two conditions on negation, the abstract counterparts of the principles
of Excluded Middle and Non-Contradiction. This *axiomatic approach* to sentential logic was later abandoned by Tarski himself, and it was not followed

by many scholars; only a few papers such as Grzegorczyk [1972], Pogorzelski and Słupecki [1960a], [1960b] and Porębska and Wroński [1975] present similar characterizations of, mainly, intuitionistic logic and some of its usual fragments. The properties involved in such characterizations are called *Tarski-style conditions* in Wójcicki [1988] (see its Section 2.3 for a discussion, which also touches on the connection of these issues with rules of Natural Deduction and Gentzen calculi); for broader accounts of Tarski's own contributions, see Blok and Pigozzi [1988] and Czelakowski and Malinowski [1985]. On the other hand, a great deal of algebraic study of sentential logics, understood as structural closure operators on the algebra of formulas, has been done by many researchers (most of them Polish, but not all), the main algebraic tool being the notion of logical matrix, and a deep universal-algebraic theory has been produced; the monographs Czelakowski [1980], [1992], Pogorzelski and Wojtylak [1982], Rasiowa [1974] and Wójcicki [1984], [1988] are good accounts of parts of this work. Later and fundamental contributions to this field are Blok and Pigozzi's [1986], [1991], [1992], as will be their long-awaited papers [1989b], [200x] on the Deduction Theorem and Abstract Algebraic Logic; most of this material appears in Czelakowski's book [2001a].

To be historically accurate one should mention Smiley's discussion in pp. 433–435 of his [1962], where he shows the insufficiency of ordinary matrices to model some logics, and proposes the use of algebras with a closure operator in order to model the deducibility relation rather than theoremhood. Smiley's proposal, briefly followed in Harrop [1965], [1968], was also put forward in Makinson [1977], but apart from this it did not attract any attention from the algebraic logic community: the matrices used in Shoesmith and Smiley [1978] are the ordinary ones, and Wójcicki did not further develop the first completeness results on generalized matrices he obtained in his [1969], [1970].

Closure operators on arbitrary algebras were first used in their full force, in an attempt to build a kind of algebraic semantics for sentential logics qualitatively different from the usual one, in Brown's dissertation [1969], where the principal advisor was Suszko, and then in Bloom and Brown [1973] and Brown and Suszko [1973], published in the same booklet together with an interesting preface by Suszko; while Brown and Suszko [1973] presents the general theory with short examples, in Bloom and Brown [1973] the abstract logics consisting of a Boolean algebra and the closure operator determined by its filters are characterized, roughly speaking, by the same metalogical properties that determine classical logic, namely finitarity, the Deduction Theorem and having all the classical tautologies as theorems. Similar characterizations were obtained in Bloom [1977] for several fragments of intuitionistic logic containing conjunction in relation with the corresponding classes of algebras and their filters.

It was this last line of research that was originally followed in Barcelona, starting with Verdú's dissertation [1978], and later on by several of his fellow colleagues and their students. In his papers [1979] – [1987] he characterizes the closure operators associated with several classes of algebras in similar, natural and logically motivated ways, and conversely he shows that the existence of such abstract logics characterizes the classes of algebras involved; they are mainly lattice-like structures or implicative structures (Hilbert and Heyting algebras, etc.). These studies were extended to other classes of structures related to several modal logics (Font [1980], Font and Verdú [1979], [1989b], Jansana [1991], [1992], [1995]), three- and four-valued logics (Font [1997], Font and Rius [1990], [2000], Font and Verdú [1988], [1989a], Rius [1992]), relevance logics (Font and Rodríguez [1994], Rodríguez [1990]), and to logics associated with cardinality restrictions on the Deduction Theorem (García Lapresta [1988a], [1988b], [1991]). One of the typical kinds of results obtained in those papers is: An algebra belongs to some class **K** if and only if there is a closure operator C on its universe satisfying such and such properties (normally including finitarity) and such that $C(\{a\}) = C(\{b\})$ implies $a = b$. At the same time, in many cases it was also found that a lattice isomorphism exists, for each algebra of suitable type, between the set of closure operators on it satisfying those properties and the set of congruences of that algebra which give a quotient in the class **K** (many in the unpublished Verdú [1986] and also in Font [1987], Font and Verdú [1989b], [1991], Jansana [1995], Rius [1992], Rodríguez [1990]; for some more details see Font [1993]). These *isomorphism theorems* were regarded as a natural extension of the well-known isomorphisms found by Czelakowski, Rasiowa, Monteiro and others in many structures of implicative character (i.e., isomorphisms between congruences and subsets of some kind), which in turn generalize the well-known isomorphism between filters and congruences in Boolean algebras. Indeed, Czelakowski, just before proving Theorem II.2.10 of his [1981], says that it "generalizes some observations made independently by several people". Note that in Rasiowa [1974] the isomorphisms are not explicitly stated, but follow easily from the correspondences between filters and congruences there established. Similar results can be found in many different papers studying algebraic structures associated in some way with logic.

Although the connection with a sentential logic (where this term has the precise meaning given in Chapter 1) was clear (maybe less clear in the cases without implication), initially it was not made explicit; it happened that the "such and such properties" were always some of the key metalogical properties of the logical system associated with the class of algebras, but only in a few cases was there a proof in the literature that these properties really characterize the sentential logic (in the sense that its consequence operator is the weakest one satisfying them). After the appearance of Blok and Pigozzi [1986], [1989a], these connections began to be made explicit, and this line of work shifted its focus to presenting the classes of abstract logics under study as being naturally associated with a logic, and to derive from this a natural association between the sentential logic and a class of algebras, but a general framework to explain these associations was still lacking.

The first published paper that performs this shift is Font and Verdú [1991], where the $\{\wedge, \vee\}$-fragment of classical sentential logic is studied. There are obvious associations between this fragment and the class of distributive lattices: the class of distributive lattices is the variety generated by the two-element lattice, this lattice semantically determines the logic, and the variety is also generated by the Lindenbaum-Tarski algebra of the logic; as a consequence, equations true in the variety correspond to pairs of interderivable formulas of the logic, and quasi-equations to rules. However, in Font, Guzmán, and Verdú [1991] it was discovered that the algebra reducts of the reduced matrices for that fragment form a much smaller class, and in Font and Verdú [1991] Proposition 2.8, it is proved that the fragment is not even protoalgebraic (in the sense of Blok and Pigozzi [1986]), so that its matrix semantics does not have a good behaviour. Thus it seemed that the classical approaches do not allow a smooth expression of the relationship between this fragment and the class of distributive lattices. On the other hand, a general notion of "model of a Gentzen calculus" was presented in Font and Verdú [1991], and it was proved that there is an equivalence between the models of a natural Gentzen calculus for that fragment and the abstract logics called "distributive" (see Section 5.1.1); as a result the class of distributive lattices was shown to be exactly the class of algebra reducts of the reduced models.

These ideas opened up a new trend in Algebraic Logic, that of studying abstract logics specifically as models of Gentzen calculi, when the latter are understood as defining a consequence operation in the set of sequents of some sentential language. This line of research seems very promising, both in its extension to other logics (see Adillon and Verdú [1996], Font [1997], Font and Rius [2000], Font and Rodríguez [1994], Gil [1996], Gil, Torrens, and Verdú [1997] and Rebagliato and Verdú [1993]), and in the obtaining of a general theory of models of Gentzen systems[3] and of their algebraization, started in Rebagliato and

---

[3]The models of Gentzen systems have been used for proof-theoretic purposes in Belardinelli, Jipsen, and Ono [2004] and Galatos, Jipsen, Kowalski, and Ono [2007], and the related notion of a fully adequate Gentzen system is further studied in Font, Jansana, and Pigozzi [2001], [2006].

Verdú [1995]⁴. Moreover, these new general theories have given rise to still more general studies of the model theory of equality-free logic, as in Casanovas, Dellunde, and Jansana [1996], Dellunde [1996], Dellunde and Jansana [1996], Elgueta [1994]⁵, and to the extension to this framework of the ideas of algebraizability under the guise of "structural equivalence" between theories as in Dellunde and Jansana [1994]⁶.

At about the same time, the second author of this monograph, in an attempt to find a common setting for all isomorphism theorems already obtained, introduced in 1991 the notions of $\mathcal{S}$-algebra and of full model of an arbitrary sentential logic $\mathcal{S}$, and proved the general version included in this monograph as Theorem 2.30; soon afterwards we realized that these notions might be used to build a general framework for describing the association between a sentential logic, a class of algebras, and a class of abstract logics, in such a way that many old results become particular cases of general properties which are now seen to hold for arbitrary sentential logics. The present monograph is the first result of our investigations; some of them were already advanced in Font [1993], and a summary was presented in Font and Jansana [1995].

What is a logic?

Every proposal of a scientific theory that aims for a reasonable degree of generality must first provide an answer to a preliminary methodological question: What should its basic objects of study be? In the case of Sentential Logic, several answers can be found in the literature: For some, a logic is a set of formulas (probably closed under substitutions and other rules), while for others it is a relation of consequence among formulas (in both cases, defined either semantically or syntactically); but for others, a logic is a "calculus", either of a "Hilbert style" or of a "Gentzen style", or of some other kind of formalism, while some think that a logic should necessarily incorporate both a calculus and a semantics; for others, forcing the meaning of the word slightly outside its natural scope, a logic is just an algebra, or a truth-table. This Introduction seems to be a good place to declare our views, which of course will be reflected in our technical treatment of the subject.

---

⁴And continued in Pynko [1999] and Raftery [2006].

⁵Later publications on model theory of equality-free languages, directly or indirectly inspired by these, are Dellunde [1999], [2000a], [2000b], [2003], Elgueta [1997], [1998], Elgueta and Jansana [1999] and Keisler and Miller [2001].

⁶An even more abstract study of the idea of equivalence of consequence operators through structural translations has been started in Blok and Jónsson [2006].

We entirely agree that the study of all the issues just mentioned belongs to *Logic* as a scientific discipline; but when faced with the question of what *a logic* is, we prefer a more neutral view that sees Logic as the study of the notion of formal logical consequence; accordingly, a sentential logic is for us just a structural consequence relation (or consequence operation) on the algebra of sentential formulas. Thus, this notion includes logics defined semantically (either by logical matrices, by classes of logical matrices, or by using the ordering relation on some set, or by Kripke models, etc.) or syntactically by some kind of formal system, of which many varieties exist, including those defined implicitly as "the weakest logic satisfying such and such properties" (whenever it exists); our treatment of logics is independent of the way they are defined. Moreover, this notion of logic allows us to treat as distinct objects but on an equal footing the two notions of consequence one can associate with a "normal modal logic", one with the full Rule of Necessitation, the other one with this rule only for theorems, see Section 5.3.

In this monograph we restrict our attention to *finitary* logics, and accordingly we will use the terms *logic* and *sentential logic* to mean a finitary and structural closure operator on the algebra of sentential formulas; see page 25 for details. However, most of the results can be generalized to non-finitary sentential logics.

On the negative side, however, our choice has at least two limitations: First, for some "logical systems", usually of philosophical origin, like Relevance logics, only the formalization of a set of "theorems" is initially introduced from the external motivations, while it is not at all clear which notion of "inference" should correspond to them under the same motivations. In these cases, our results apply only, and separately, to each of the consequence relations that can be ascribed to these logical systems, and not directly to the original formalization; see for instance our treatment of Relevance Logic in Section 5.4.1. Second, it excludes from our scope the host of so-called "substructural logics" (see the foundational volume Došen and Schroeder-Heister [1993]) and other "logical systems", like non-monotonic logics, which are being studied because of their relevance to Theoretical Computer Science and other disciplines connected with the study of reasoning in (semi-)intelligent systems. Such new developments have activated debate about the very question of *what is a logical system?*, as witnessed by the collection Gabbay [1994].

Outline of the contents

Chapter 1 collects the preliminary definitions and notations concerning logical matrices, abstract logics and sentential logics, and contains the portion of the general theory of abstract logics needed in the rest of the monograph. In this chapter we have included results already obtained in Brown and Suszko [1973] and in Verdú [1978], [1987], together with new ones, forming a unified exposition of (a fragment of) the partly unpublished "folklore" of the field. Although we give references for some definitions or results, they should not be taken as historical attributions, but rather as notifications of other places where more details can be found.

The main tool of the monograph will be the notion of the *Tarski congruence* $\widetilde{\Omega}(\mathbb{L})$ associated with an abstract logic $\mathbb{L} = \langle A, C \rangle$; it is the greatest congruence of the algebra A which is compatible with the abstract logic $\mathbb{L}$, i.e., which does not identify elements with different closure (Definition 1.1). This defines on every algebra A the *Tarski operator* $\widetilde{\Omega}_A$ which assigns to every abstract logic $\mathbb{L} = \langle A, C \rangle$ over the algebra A its Tarski congruence $\widetilde{\Omega}(\mathbb{L})$. These notions are, in some sense, extensions of the notions of Leibniz congruence and Leibniz operator due to Blok and Pigozzi, and are the generalization of the procedure usually followed in the literature, and particularly by Tarski, when the so-called *Lindenbaum-Tarski algebra* of a sentential logic is constructed (for more details see pages 19 and 29). Several of its properties will also be, to a certain extent, a generalization of some properties of the Leibniz operator of algebraizable or protoalgebraic logics; in this chapter the most elementary ones are presented, especially those dealing with the process of *reduction* of an abstract logic, which consists in factoring an abstract logic by its Tarski congruence. An abstract logic is *reduced* when its Tarski congruence is the identity. The few results we need on logical congruences, quotients and homomorphisms, parallel to well-known facts of universal algebra, are also presented.

Chapter 2 contains the definition of the notions of S-algebra and of full model of an arbitrary sentential logic S, and the study of their general properties. It starts (Section 2.1) from the consideration of abstract logics as *models* of sentential logics, in a completely natural way (which amounts to being a *generalized matrix* in the sense of Wójcicki), and we select the *full models* as those such that their reduction has as closed sets all the filters of the sentential logic on the quotient algebra. In Section 2.2 the S-algebras are introduced as the algebraic reducts of the reduced full models of the logic, and several properties of the class **Alg**S of all the S-algebras are proved. We highlight the Completeness Theorem 2.22 and Theorem 2.23 stating that **Alg**S is the class of all subdirect products of members of the class of algebraic reducts of reduced matrices of the logic; from this fact some sufficient conditions for the coincidence of both classes of algebras are derived. Section 2.3 is mainly devoted to the proof of the central Theorem 2.30, stating that for every algebra A, the Tarski operator $\widetilde{\Omega}_A$ is an isomorphism between the ordered sets of all the full models of S on A and all the congruences of A whose quotient algebra belongs to the class **Alg**S. This isomorphism, which results in a lattice isomorphism, is, in some sense, an extension of one part of Theorem 5.1 of Blok and Pigozzi [1989a], which establishes (for an algebraizable logic S) that the Leibniz operator on every algebra A is a lattice isomorphism between the S-filters on A and the congruences of A whose quotient belongs to the equivalent quasivariety semantics of S; but at the same time, as we have already said, Theorem 2.30 is the general property corresponding to many particular cases proved by Verdú and others. This section also contains some categorial formulations of the equivalence between S-algebras and reduced full models, and of the fact that the process of reduction can be seen as a reflector from the category of all full models to the full subcategory of the reduced ones. Finally Section 2.4 begins the study of how metalogical properties of a sentential logic are "inherited" by all its full models, an issue underlying many of our intuitions. It is proved that some properties, like the Deduction Theorem, the Properties of Conjunction and Disjunction, and the Introduction of a modal operator, pass from a sentential logic to all its full models, while others, like the Reductio ad Absurdum, do not. Some attention is devoted to the Congruence Property (that the interderivability relation is a congruence of all the connectives of the logic). Logics having this property have been called *selfextensional*, and we call *strongly selfextensional*[7] those whose full models all have it. While it is still an open question whether there is a selfextensional sentential logic that is not strongly selfextensional, as an application of the results of Chapter 4 we are able to see that the answer is negative for logics with Conjunction and for logics having a certain form of the Deduction Theorem[8].

In Chapter 3 we apply the notions and results of the previous chapter to find the S-algebras and the full models of sentential logics which are protoalgebraic or algebraizable. We prove that in such a case the class of S-algebras is exactly the class of algebras ordinarily associated with the logic, i.e., the class of algebraic reducts of reduced matrices, or the equivalent quasivariety semantics for the algebraizable logics. One of the themes of this chapter is the relationship between full models of S and the abstract logics whose closure system consists of all the S-filters containing a fixed one. We prove that a logic is protoalgebraic iff all its full models have this form (Theorem 3.4), characterize the S-filters which are theorems of a full model, and obtain a new and interesting class of sentential logics: those where this correspondence establishes a complete identification between S-filters and full models; Theorem 3.8 contains several characterizations of

---

[7]Since this is a property of the class of full models of a logic, in later publications the alternative, more descriptive term *fully selfextensional* has been adopted.

[8]The above question has been answered in the affirmative in Babyonyshev [2003].

this interesting class of sentential logics, called *weakly algebraizable*. The logics in this class have the outstanding property that the Leibniz operator establishes an isomorphism between $\mathcal{S}$-filters and congruences whose quotient belongs to **Alg$\mathcal{S}$**, a property that characterizes algebraizable logics when the class **Alg$\mathcal{S}$** is a quasi-variety. We obtain other interesting characterizations of algebraizable logics. The same theme restricted to full models on the formula algebra leads us to consider the so-called *Fregean* logics (those where the interderivability relation modulo an arbitrary theory of the logic is a congruence), and the *Fregean protoalgebraic* logics, already studied by Pigozzi and Czelakowski. As an application of our results we obtain a new proof (Theorem 3.18) of the result, already found by them in a different context[9], that every Fregean protoalgebraic logic with theorems is regularly algebraizable. The chapter closes with the proof (Corollary 3.21) that if a logic is weakly algebraizable then it is strongly selfextensional if and only if it is Fregean. This and other results clarify to some extent the topography of the logics around these properties.

The notion of full model seems to be inherently of higher order nature; therefore it seems interesting to try to characterize it in a more practical way. Using essentially Proposition 2.21 we can see (and this is done in detail in Chapter 5) that many old results are characterizations of the full models of some sentential logics as those abstract logics satisfying certain properties concerning the relationship between the closure operator and the operations of the algebra, properties which are metalogical properties of the sentential logic. A large and important class of metalogical properties of a sentential logic are those expressible as a *Gentzen-style rule*, i.e., as a rule of some Gentzen system. So there arises the question of whether we can always describe the full models of a sentential logic as the models of some set of Gentzen-style rules. We treat this issue more generally in Chapter 4. Section 4.1 contains all general definitions and results, including that of a Gentzen system, the notion of model of a Gentzen system (a natural use of abstract logics, at least for Gentzen systems with structural rules), and that of a Gentzen system being *strongly adequate*[10] for a sentential logic: Roughly speaking, this happens when the full models of the sentential logic are exactly the finitary models of the Gentzen system. This relationship between a Gentzen system and a sentential logic is very strong: although not every sentential logic has a strongly adequate Gentzen system, if it exists then it is unique and the full models of the sentential logic can be described by the rules of the Gentzen system; in particular, in this situation the $\mathcal{S}$-algebras are the algebraic reducts of the reduced

---

[9] See Section 6.2 of Czelakowski [2001a].

[10] Again, since this is a property related to the class of full models of a logic, in later publications the more descriptive term *fully adequate* has been adopted.

models of the Gentzen system. The use we make of Gentzen systems leads us to a point of contact with a different and very recent trend in Algebraic Logic, that of the *algebraization of Gentzen systems*, started in Rebagliato and Verdú [1993], [1995]. We find a situation where the result of the algebraization of a sentential logic found through that of a Gentzen system related to it completely agrees with the algebraization we find with our notions. Sections 4.2 and 4.3 treat in parallel the cases of selfextensional logics with Conjunction and with the Deduction Theorem, respectively. We associate a Gentzen system in a canonical way with each logic in one of these classes, prove that it is algebraizable in the sense of Rebagliato and Verdú [1993], [1995], and that the corresponding class of algebras is the variety generated by the Lindenbaum-Tarski algebra of the sentential logic. Using this fact we show that the Gentzen system is strongly adequate for the logic, and that the logic is strongly selfextensional; therefore the Congruence Property is inherited by all the full models. As a by-product we obtain the result that every Fregean protoalgebraic logic with Conjunction or with the Deduction Theorem is *strongly algebraizable* (i.e., it is algebraizable and the equivalent quasivariety semantics is in fact a variety); these results have been obtained by Czelakowski and Pigozzi using a different framework[11].

Finally Chapter 5 applies all the preceding methods and results to the study of particular sentential logics. Wherever possible we have classified them according to the definitions given in the monograph; as a result we have found counterexamples to several questions raised in the text. We determine the classes of S-algebras and of full models of a number of sentential logics, either by just putting together already published results on abstract logics and some of the general results contained in the preceding chapters, or by showing in more detail how the proof proceeds, using if necessary published or unpublished material on the logics under consideration. Of special interest are, of course, the non-protoalgebraic cases, but even for the protoalgebraic cases this study is interesting, since among them the non-algebraizable cases cannot always be distinguished by their S-algebras; indeed, in Sections 5.3 and 5.4 we present a number of examples of pairs of sentential logics (of which one is algebraizable and the other is not) sharing the same class of S-algebras, but with different full models. This chapter draws attention to the need for a thorough investigation of a larger number of sentential logics in the light of our approach, particularly finding the S-algebras and the full models of many of the non-algebraizable ones.

---

[11] These results have been finally published in Czelakowski and Pigozzi [2004a], [2004b]; the treatment in the first of these papers incorporates several aspects and techniques introduced in the present monograph.

This monograph is the first detailed exposition of our theory. As is to be expected, there is plenty of room in it for further research. Specifically we have highlighted several *open problems* at different places in the text.

Acknowledgements

Our main scientific debts concerning this monograph are to Ventura Verdú, Wim Blok, Don Pigozzi and Janusz Czelakowski. We particularly acknowledge Don and Janusz for their interest and patience in reading a first version; their comments, questions and suggestions for further work contributed (in some cases decisively) to improving the contents and the exposition. Our work is part of the research project in *Algebraic Logic* of the Barcelona group, which was partially supported by grants PB90-0465-C02-01 and PB94-0920 of the Spanish DGICYT. Some diagrams were drawn using Paul Taylor's macros. Finally, we want to thank also Donald Knuth and Leslie Lamport for having created these excellent tools for the working mathematician, TEX and LATEX. Without their existence this work would have never reached its present (printed) stage; we entirely agree that, paraphrasing what Paul Halmos once said (see Knuth, Larrabee, and Roberts [1989], p. 106),

> "mathematicians who merely *think* their theorems
> have no more done their job
> than painters who merely *think* their paintings".

Note to the second edition (2009)

We have corrected all the typos and mistakes found to date, as well as a few minor inaccuracies of exposition. We have updated all bibliographical references to items quoted as "to appear" in 1996; this explains that some of them have a later date (and, in some cases, a different title). We have adopted the ASL recommended "author-year" style of citations, which has implied small adjustments in some of the sentences containing them. No further changes have been made to the real contents of the monograph.

However, since the subject (now called *abstract algebraic logic*) has been naturally growing and evolving over the time in several directions, we have added a number of footnotes (the few ones in the first edition have been moved to the main text, hence all present footnotes correspond to the second edition) informing about major advances, solved open problems, new relevant publications, changes in terminology, and so on. As general sources of survey-style information on later developments in the field, see Font, Jansana, and Pigozzi [2003] and Font [2003b], [2006].

Our work on this second edition has been supported by grants MTM2008-01139 from the Spanish government (which includes EU's FEDER funds) and 2005SGR-00083 from the Catalan government.

CHAPTER 1

GENERALITIES ON ABSTRACT LOGICS
AND SENTENTIAL LOGICS

In this chapter we include the main definitions, notations, and general properties concerning logical matrices, abstract logics and sentential logics. Most of the
results reproduced here are not new; however, those concerning abstract logics
are not well-known, so it seems useful to recall them in some detail, and to prove
some of the ones that are new. Useful references on these topics are Brown and
Suszko [1973], Burris and Sankappanavar [1981] and Wójcicki [1988].

Algebras

In this monograph (except in Chapter 5, where we deal with examples) we
will always work with algebras $A = \langle A, \dots \rangle$ of the same, arbitrary, similarity
type; thus, when we say "every/any/some algebra" we mean "of the same type".
By $\mathrm{Hom}(A, B)$ we denote the set of all *homomorphisms* from the algebra A
into the algebra B. The set of *congruences* of the algebra A will be denoted by
$\mathrm{Con}\, A$. Many of the sets we will consider have the structure of a (often complete,
or even algebraic) lattice, but we will not use a different symbol for the lattice
and for the underlying set, since no confusion is likely to arise. Given any class
$\mathbf{K}$ of algebras, the set $\mathrm{Con}_{\mathbf{K}} A = \{\theta \in \mathrm{Con}\, A : A/\theta \in \mathbf{K}\}$ is called the set
of $\mathbf{K}$-*congruences* of A; while this set is ordered under $\subseteq$, in general it is not a
lattice. This set will play an important role in this monograph.

Formulas, equations, interpretations

We will denote by $\boldsymbol{Fm} = \langle Fm, \dots \rangle$ the absolutely free algebra of the similarity type under consideration generated by some fixed but unspecified set Var,
which we assume to be countably infinite. $\boldsymbol{Fm}$ is usually called the *algebra of
formulas* (or the algebra of terms), and the elements of Var the *variables*, or
atomic formulas. The letters $p, q, \dots$ will denote variables, and the formulas will
be denoted by lowercase Greek letters such as $\varphi, \psi, \xi, \eta, \dots$, while uppercase

Greek letters such as Γ, Δ will denote sets of formulas. The **equations** of the similarity type are pairs $\langle \varphi, \psi \rangle$ of formulas, which we write as $\varphi \approx \psi$; the set of all equations will be denoted by $\mathrm{Eq}(\boldsymbol{Fm})$.

Let $\boldsymbol{A}$ be any algebra of the same similarity type as $\boldsymbol{Fm}$. An **interpretation** in $\boldsymbol{A}$ is any $h \in \mathrm{Hom}(\boldsymbol{Fm}, \boldsymbol{A})$, that is, any homomorphism (in the ordinary, algebraic, sense) from $\boldsymbol{Fm}$ into $\boldsymbol{A}$; because of the freeness of $\boldsymbol{Fm}$ any such interpretation is completely determined by its restriction to Var. Therefore for any $\varphi \in Fm$, the value $h(\varphi)$ is determined by the values of those $p \in Var$ that appear in φ. We will often use the convention of writing $\varphi(p, q, r, \dots)$ to mean that the variables appearing in φ are among those in $\{p, q, r, \dots\}$; then given elements $a, b, c, \dots \in A$ we put $\varphi^{\boldsymbol{A}}(a, b, c, \dots)$ for $h(\varphi)$ whenever $h(p) = a, h(q) = b, h(r) = c, \dots$; we will also use the vectorial notations $\vec{q}$ and $\vec{a}$ for sequences of variables and of elements of A, and write $\varphi(\vec{q})$ and $\varphi^{\boldsymbol{A}}(\vec{a})$, respectively. These conventions are extended to sets of formulas: if $\Gamma \subseteq Fm$ then $\Gamma^{\boldsymbol{A}}(\vec{a})$ stands for $h[\Gamma]$ where h is any interpretation such that $h(p_i) = a_i$, and the variables appearing in Γ are among the p_i. A **substitution** is any homomorphism from the formula algebra into itself.

Matrices

A **matrix** or *logical matrix* is a pair $\mathcal{M} = \langle \boldsymbol{A}, F \rangle$ where $\boldsymbol{A}$ is an algebra and $F \subseteq A$; F is sometimes referred to as the **filter** of the matrix. Given any $\theta \in \mathrm{Con}\,\boldsymbol{A}$ we can construct the **quotient matrix** $\mathcal{M}/\theta = \langle \boldsymbol{A}/\theta, F/\theta \rangle$, where $\boldsymbol{A}/\theta$ is the ordinary quotient algebra and $F/\theta = \{a/\theta : a \in F\}$. Making this quotient is reasonable in this context only when $\theta \in \mathrm{Con}\,\boldsymbol{A}$ is **compatible** with F: This means that for any $\langle a, b \rangle \in \theta$ it happens that $a \in F$ if and only if $b \in F$, that is, θ does not identify elements inside F with elements outside F; in such a case one also says that θ is a **matrix congruence** of $\mathcal{M}$; the set $\mathrm{Con}\,\mathcal{M}$ of all these congruences is a principal ideal (and hence a sublattice) of the lattice $\mathrm{Con}\,\boldsymbol{A}$; its maximum element is called the **Leibniz congruence** of the matrix, and is denoted by $\boldsymbol{\Omega_A}(F)$. The reason for naming it after Leibniz is clearly explained by its inventors Blok and Pigozzi in their [1989a] pp. 10–11, and is related to the following characterization (see Czelakowski [1980] Theorem 3.2, and also Wójcicki [1988] Lemma 3.1.10): If $a, b \in A$, then

$$
\begin{aligned}
\langle a, b \rangle \in \boldsymbol{\Omega_A}(F) &\iff \forall \varphi(p, \vec{q}) \in Fm\,, \ \forall \vec{c} \in A^k\,, \\
\varphi^{\boldsymbol{A}}(a, \vec{c}) \in F &\iff \varphi^{\boldsymbol{A}}(b, \vec{c}) \in F\,.
\end{aligned}
\tag{1.1}
$$

The natural number k is the length of $\vec{q}$; it obviously depends on φ.

The mapping $F \mapsto \Omega_A(F)$ is called the *Leibniz operator* of the algebra A. A matrix is *reduced* when its only matrix congruence is the identity relation, that is, $\Omega_A(F) = Id_A$. For any matrix $\mathcal{M} = \langle A, F \rangle$, the *quotient matrix* $\mathcal{M}^* = \mathcal{M}/\Omega_A(F) = \langle A/\Omega_A(F), F/\Omega_A(F) \rangle$ is reduced, and is called the *reduction* of $\mathcal{M}$. Given any class of matrices **M**, we put $\mathbf{M}^* = \{\mathcal{M}^* : \mathcal{M} \in \mathbf{M}\}$.

Abstract logics

By a *closure operator* on a set A we mean, as is usual, a mapping $C : P(A) \to P(A)$, where $P(A)$ is the power set of A, such that for all $X, Y \subseteq A$,

(C1) $X \subseteq C(X)$,
(C2) If $X \subseteq Y$ then $C(X) \subseteq C(Y)$, and
(C3) $C(C(X)) = C(X)$.

By a *closure system* on a set A we understand a family $\mathcal{C}$ of subsets of A such that $A \in \mathcal{C}$ and $\mathcal{C}$ is closed under arbitrary intersections. Given a closure operator C on A, the family $\mathcal{C} = \{X \subseteq A : C(X) = X\}$ of its *closed* sets is a closure system, and conversely given a closure system $\mathcal{C}$ on A, the function defined by $X \longmapsto C(X) = \bigcap\{T \in \mathcal{C} : X \subseteq T\}$ is a closure operator; and these two correspondences are inverse to one another. A closure operator is *finitary* whenever it satisfies $C(X) = \bigcup\{C(F) : F \subseteq X, F \text{ finite}\}$ for any $X \subseteq A$; an equivalent statement is that the closure system $\mathcal{C}$ is *inductive*, i.e., closed under unions of upwards directed subfamilies (the union of the empty family is taken to be A).

An *abstract logic* is a pair $\mathbb{L} = \langle A, C \rangle$ where A is an algebra, and C is a closure operator on A. The elements of $C(\emptyset)$ are called the *theorems* of $\mathbb{L}$. With any abstract logic we associate the closure system $\mathcal{C} = Th\mathbb{L}$ of its closed sets (also called *theories*); given the duality existing between closure operators and closure systems, we will also present abstract logics as pairs $\mathbb{L} = \langle A, \mathcal{C} \rangle$ where $\mathcal{C}$ is a closure system on A. Some kind of "typographical correspondence" between pairs of associated closure operators and closure systems, like $C \cdots \mathcal{C}$, $F \cdots \mathcal{F}$, etc., will be used without notification; likewise, when super- or subscripting an abstract logic, we will suppose that, unless otherwise specified, the super- or subscripts are also applied to the corresponding algebra, closure operator and closure system. Sometimes it will be convenient to write $\mathbb{L} = \langle A_{\mathbb{L}}, C_{\mathbb{L}} \rangle$ and $\mathbb{L} = \langle A_{\mathbb{L}}, \mathcal{C}_{\mathbb{L}} \rangle$. We use the customary abbreviations $C(a)$ for $C(\{a\})$, $C(X, a)$ for $C(X \cup \{a\})$ and so on.

It will be useful to remember that all closure systems are complete lattices (where the infimum of any family of closed sets is its intersection while its supremum is the closure of its union), and that any complete lattice is isomorphic to a closure system; see for instance Burris and Sankappanavar [1981] Section I.5.

Abstract logics on the same algebra are ordered according to the corresponding closure operators: We say that $\mathbb{L}$ is **weaker** than $\mathbb{L}'$, and that $\mathbb{L}'$ is **stronger** than $\mathbb{L}$, in symbols $\mathbb{L} \leqslant \mathbb{L}'$, when $\mathrm{C} \leqslant \mathrm{C}'$, that is, when for any $X \subseteq A$, $\mathrm{C}(X) \subseteq \mathrm{C}'(X)$; this is equivalent to the reverse order among closure systems: $\mathbb{L} \leqslant \mathbb{L}'$ iff $\mathcal{C}' \subseteq \mathcal{C}$. In case $\mathcal{C}' \subseteq \mathcal{C}$ we say that $\mathcal{C}$ is **finer** than $\mathcal{C}'$. It is easy to see that the set of all abstract logics on the same algebra A equipped with this ordering is a complete lattice, dually isomorphic to the complete lattice of all closure systems on A ordered under $\subseteq$. When $\mathbb{L} \leqslant \mathbb{L}'$ we also say that $\mathbb{L}'$ is an **extension** of $\mathbb{L}$.

If $\mathcal{C}$ is a closure system on A then for any $T \subseteq A$, the family of all closed sets containing T is also a closure system, denoted by $\mathcal{C}^T = \{S \in \mathcal{C} : T \subseteq S\}$. We will often use this construction, which associates with any abstract logic $\mathbb{L}$ and any $T \subseteq A$ the abstract logic $\mathbb{L}^T = \langle A, \mathcal{C}^T \rangle$ or $\langle A, \mathrm{C}^T \rangle$, called the **axiomatic extension** of $\mathbb{L}$ by T; since for any $X \subseteq A$, $\mathrm{C}^T(X) = \mathrm{C}(T \cup X)$, this extension is the same for all $T \subseteq A$ having the same closure under C, and we often restrict its use to the $T \in \mathcal{C}$.

With any abstract logic $\mathbb{L} = \langle A, \mathrm{C} \rangle$ we can associate the family or "bundle" of matrices $\{\langle A, T \rangle : T \in \mathcal{C}\}$. Conversely, any bundle of matrices having the same algebra reduct originates an abstract logic, whose closure system is generated by the family of filters of the matrices in the bundle. Bundles of matrices have sometimes been referred to also as *generalized matrices*; see Wójcicki [1973], and also our Proposition 2.7 and subsequent comments.

Logical congruences

If $\mathbb{L} = \langle A, \mathrm{C} \rangle$ is an abstract logic, then a congruence $\theta \in \mathrm{Con}\, A$ is a **logical congruence** of $\mathbb{L}$ when $\langle a, b \rangle \in \theta$ implies $\mathrm{C}(a) = \mathrm{C}(b)$; or, equivalently, when θ is compatible with every $T \in \mathcal{C}$. We denote by $\mathrm{Con}\,\mathbb{L}$ the set of all logical congruences of $\mathbb{L}$; from the preceding observation it follows that

$$\mathrm{Con}\,\mathbb{L} = \bigcap_{T \in \mathcal{C}} \mathrm{Con}\,\langle A, T \rangle. \qquad (1.2)$$

It is easy to see that this set, ordered by $\subseteq$, is a complete lattice, and a principal ideal of the lattice $\mathrm{Con}\,A$. Actually, its generator turns out to be the most important tool in our theory:

DEFINITION 1.1. *If* $\mathbb{L} = \langle A, \mathrm{C} \rangle$ *is an abstract logic, the **Tarski congruence** of* $\mathbb{L}$ *is*

$$\widetilde{\Omega}(\mathbb{L}) = \max \mathrm{Con}\,\mathbb{L},$$

i.e., the greatest logical congruence of $\mathbb{L}$. *For every algebra* A, *the* **Tarski operator** *on* A *is the mapping*

$$\widetilde{\Omega}_A : \mathbb{L} \longmapsto \widetilde{\Omega}(\mathbb{L}),$$

that is, the mapping $\mathbb{L} \mapsto \widetilde{\Omega}(\mathbb{L})$ *restricted to abstract logics on* A.

Given an algebra A, for every closure operator C on A we can consider the abstract logic $\mathbb{L} = \langle A, \mathrm{C} \rangle$; therefore it is natural to extend the notations introduced above and write $\widetilde{\Omega}_A(\mathrm{C})$ for $\widetilde{\Omega}(\langle A, \mathrm{C} \rangle)$; similarly, if $\mathcal{C}$ is a closure system on A, we write $\widetilde{\Omega}_A(\mathcal{C})$ for $\widetilde{\Omega}(\langle A, \mathcal{C} \rangle)$. The mapping $\mathrm{C} \mapsto \widetilde{\Omega}_A(\mathrm{C})$ will be identified, for practical purposes, with the Tarski operator on A. From the definition it follows that $\mathrm{Con}\,\mathbb{L} = \{ \theta \in \mathrm{Con}\,A : \theta \subseteq \widetilde{\Omega}(\mathbb{L}) \}$; moreover one can prove, using (1.2), the following:

PROPOSITION 1.2. *For any abstract logic* $\mathbb{L}$, $\widetilde{\Omega}(\mathbb{L}) = \bigcap \{ \Omega_A(T) : T \in \mathcal{C}_{\mathbb{L}} \}$; *that is,* $\widetilde{\Omega}_A(\mathcal{C}) = \bigcap \{ \Omega_A(T) : T \in \mathcal{C} \}$ *for any algebra* A *and any closure system* $\mathcal{C}$ *on* A.   ⊣

As a consequence of this and of (1.1), it follows that for any abstract logic $\mathbb{L} = \langle A, \mathrm{C} \rangle$, the Tarski congruence of $\mathbb{L}$ can be characterized as:

$$
\begin{aligned}
\langle a, b \rangle \in \widetilde{\Omega}(\mathbb{L}) \iff & \forall \varphi(p, \vec{q}) \in Fm \,, \forall \vec{c} \in A^k \,, \\
& \mathrm{C}\big(\varphi^A(a, \vec{c})\big) = \mathrm{C}\big(\varphi^A(b, \vec{c})\big)
\end{aligned}
\tag{1.3}
$$

Thus the notions of the Tarski congruence/operator are, in some sense, extensions of the notions of the Leibniz congruence/operator; actually they were called "extended Leibniz congruence/operator" in Font [1993], where the notions of Tarski congruence/operator were introduced. The reason for naming them after Tarski is that this relation is the one Tarski took when he factored the formula algebra of classical logic to find a Boolean algebra for the first time; in this case the relation had the particular form: $\varphi \equiv \psi \iff \vdash \varphi \leftrightarrow \psi$; it is interesting to note that the relation expressed by the Tarski congruence in the case of sentential logics (in the form of expression (1.6) of page 29) was already considered in Smiley [1962], where it is called "synonymity" and is presented as the true general notion of equivalence of formulas, of which Tarski's $\equiv$ is but a particular form suitable for classical logic (due to the Deduction Theorem).

From Proposition 1.2 follows at once:

PROPOSITION 1.3. *On every algebra* A, *the Tarski operator* $\widetilde{\Omega}_A$ *is order-preserving, in the sense that, if* $\mathbb{L}, \mathbb{L}'$ *are abstract logics on* A *such that* $\mathbb{L} \leqslant \mathbb{L}'$, *then* $\widetilde{\Omega}(\mathbb{L}) \subseteq \widetilde{\Omega}(\mathbb{L}')$.   ⊣

Bilogical morphisms and logical quotients

Given two abstract logics $\mathbb{L}$ and $\mathbb{L}'$, an homomorphism $h \in \mathrm{Hom}(A, A')$ is a *logical morphism* from $\mathbb{L}$ into $\mathbb{L}'$ when $h^{-1}[T] \in \mathcal{C}$ for any $T \in \mathcal{C}'$. An abstract logic $\mathbb{L} = \langle A, \mathcal{C} \rangle$ is said to be *projectively generated* from a family $\{\mathbb{L}_i : i \in I\}$ of abstract logics by a family of homomorphisms $\{h_i \in \mathrm{Hom}(A, A_i) : i \in I\}$ when $\mathbb{L}$ is the strongest abstract logic such that each of the h_i is a logical morphism; that is, when the closure system $\mathcal{C}$ is the smallest one including the sets $h_i^{-1}[T]$ for all $T \in \mathcal{C}_i$ and all $i \in I$. We will deal almost exclusively with the particular case where the generating families reduce to one abstract logic and one homomorphism; in this case $\mathbb{L}$ is projectively generated from $\mathbb{L}'$ by h when $\mathcal{C} = \{h^{-1}[T] : T \in \mathcal{C}'\}$. A mapping h is a *bilogical morphism from $\mathbb{L}$ onto $\mathbb{L}'$* (or *between $\mathbb{L}$ and $\mathbb{L}'$*) when it is an epimorphism and projectively generates $\mathbb{L}$ from $\mathbb{L}'$.

These notions were introduced in Brown and Suszko [1973]. The main properties of bilogical morphisms that we will need are summarized in the next two propositions; they are mainly due to Brown, Suszko and Verdú.

PROPOSITION 1.4. *Let $\mathbb{L}$ and $\mathbb{L}'$ be two abstract logics, and $h \in \mathrm{Hom}(A, A')$ be an epimorphism. Then the following properties are equivalent:*

(i) *h is a bilogical morphism between $\mathbb{L}$ and $\mathbb{L}'$.*
(ii) *$\forall X \subseteq A$, $\mathrm{C}(X) = h^{-1}[\mathrm{C}'(h[X])]$; that is, $a \in \mathrm{C}(X)$ iff $h(a) \in \mathrm{C}'(h[X])$.*
(iii) *$\forall X \subseteq A$, $h[\mathrm{C}(X)] = \mathrm{C}'(h[X])$, and $\ker h \in \mathrm{Con}\mathbb{L}$.*
(iv) *$\forall Y \subseteq A'$, $\mathrm{C}'(Y) = h[\mathrm{C}(h^{-1}[Y])]$, and $\ker h \in \mathrm{Con}\mathbb{L}$.*
(v) *$\mathcal{C}' = \{h[T] : T \in \mathcal{C}\}$ and $\ker h \in \mathrm{Con}\mathbb{L}$.*
(vi) *$\mathcal{C} = \{h^{-1}[S] : S \in \mathcal{C}'\}$.*   ⊣

Note that what the condition "$\ker h \in \mathrm{Con}\mathbb{L}$" says is just that for any $a, b \in A$, $h(a) = h(b)$ implies $\mathrm{C}(a) = \mathrm{C}(b)$. This condition is usually easily verifiable, and actually items (iii) and (iv) are two of the most useful characterizations of the notion of bilogical morphism.

PROPOSITION 1.5. *Let $\mathbb{L}$ and $\mathbb{L}'$ be two abstract logics, and $h \in \mathrm{Hom}(A, A')$ be an epimorphism. Then h is a bilogical morphism from $\mathbb{L}$ onto $\mathbb{L}'$ if and only if their lattices of theories $\mathcal{C}$ and $\mathcal{C}'$ are isomorphic under the correspondence induced by h. In particular for all $T \in \mathcal{C}$, $h^{-1}[h[T]] = T$, and for all $S \in \mathcal{C}'$, $h[h^{-1}[S]] = S$.*   ⊣

This very strong relation between the lattices of theories of two abstract logics when there is a bilogical morphism between them has several consequences:

COROLLARY 1.6. *Let $\mathbb{L}_1$ and $\mathbb{L}_2$ be abstract logics, and let h be a bilogical morphism between them. Then the mapping $\mathcal{C} \mapsto \{h[X] : X \in \mathcal{C}\}$ is an isomorphism between the lattice of all abstract logics on A_1 extending $\mathbb{L}_1$ and the lattice of all abstract logics on A_2 extending $\mathbb{L}_2$.* ⊣

In the next statement we use the following notation: If $h : A \to B$ is any mapping and $R \subseteq B \times B$, then $h^{-1}[R] = \{\langle x, y \rangle \in A \times A : \langle h(x), h(y) \rangle \in R\}$.

PROPOSITION 1.7. *Let h be a bilogical morphism between the abstract logics $\mathbb{L}_1 = \langle A_1, C_1 \rangle$ and $\mathbb{L}_2 = \langle A_2, C_2 \rangle$. Then for any $T \in \mathcal{C}_2$, $h^{-1}\big[\widetilde{\Omega}(\mathbb{L}_2^T)\big] = \widetilde{\Omega}\big(\mathbb{L}_1^{h^{-1}[T]}\big)$; in particular we have that*

$$\widetilde{\Omega}(\mathbb{L}_1) = h^{-1}\big[\widetilde{\Omega}(\mathbb{L}_2)\big].$$

PROOF. Using the characterization (1.3) of the Tarski congruence and bearing in mind that h is a bilogical morphism and so an epimorphism, it is easy to check that, for any $a, b \in A_1$,

$\langle a, b \rangle \in h^{-1}\big[\widetilde{\Omega}(\mathbb{L}_2^T)\big]$

iff $\langle h(a), h(b) \rangle \in \widetilde{\Omega}(\mathbb{L}_2^T)$

iff $\forall \varphi(p, \vec{q}), \forall \vec{c} \in A_2^k, \; C_2^T\big(\varphi^{A_2}(h(a), \vec{c})\big) = C_2^T\big(\varphi^{A_2}(h(b), \vec{c})\big)$

iff $\forall \varphi(p, \vec{q}), \forall \vec{d} \in A_1^k, \; C_2^T\big(h(\varphi^{A_1}(a, \vec{d}))\big) = C_2^T\big(h(\varphi^{A_1}(b, \vec{d}))\big)$

iff $\forall \varphi(p, \vec{q}), \forall \vec{d} \in A_1^k, \; h\big[C_1^{h^{-1}[T]}\big(\varphi^{A_1}(a, \vec{d})\big)\big] = h\big[C_1^{h^{-1}[T]}\big(\varphi^{A_1}(b, \vec{d})\big)\big]$

iff $\forall \varphi(p, \vec{q}), \forall \vec{d} \in A_1^k, \; C_1^{h^{-1}[T]}\big(\varphi^{A_1}(a, \vec{d})\big) = C_1^{h^{-1}[T]}\big(\varphi^{A_1}(b, \vec{d})\big)$

iff $\langle a, b \rangle \in \widetilde{\Omega}\big(\mathbb{L}_1^{h^{-1}[T]}\big)$.

By taking $T = C_2(\emptyset)$ we obtain the second part. ⊣

Two abstract logics $\mathbb{L}$ and $\mathbb{L}'$ are **isomorphic** (and we write $\mathbb{L} \cong \mathbb{L}'$) when there is a bijective logical morphism between them whose inverse is also a logical morphism. This is equivalent to saying that there is a bilogical morphism between them which is an isomorphism between $A_{\mathbb{L}}$ and $A_{\mathbb{L}'}$, and also to saying that there is an isomorphism between $A_{\mathbb{L}}$ and $A_{\mathbb{L}'}$ such that $\mathcal{C}_{\mathbb{L}'} = \{h[T] : T \in \mathcal{C}_{\mathbb{L}}\}$.

If $\mathbb{L} = \langle A, \mathcal{C} \rangle$ is an abstract logic and $\theta \in \operatorname{Con} A$, an abstract logic can be obtained on the quotient algebra A/θ by defining $\mathcal{C}/\theta = \{S \subseteq A/\theta : \pi_\theta^{-1}[S] \in \mathcal{C}\}$, where π_θ is the natural epimorphism or **projection** of A onto A/θ; we put $\mathbb{L}/\theta = \langle A/\theta, \mathcal{C}/\theta \rangle$, call this the **logical quotient** of $\mathbb{L}$ by θ, and denote the corresponding closure operator by C/θ. Then obviously π_θ is a logical morphism between $\mathbb{L}$ and $\mathbb{L}/\theta$. If in addition θ is a logical congruence of $\mathbb{L}$, then $\mathcal{C}/\theta = \{\pi_\theta[T] : T \in \mathcal{C}\}$ and π_θ becomes a bilogical morphism between $\mathbb{L}$ and $\mathbb{L}/\theta$.

The following results lead us to conclude that the roles of logical congruences and bilogical morphisms in the theory of abstract logics, and the relationships between them, are parallel to those of congruences and epimorphisms in universal algebra, and especially similar to those of matrix congruences and "strict" matrix epimorphisms (also called "strong", "reductive" or "contractive" in the literature) in the theory of logical matrices, as developed for instance in Blok and Pigozzi [1986], [1992], Czelakowski [1980], or Wójcicki [1988].

THEOREM 1.8 (Homomorphism Theorem). *If h is a bilogical morphism between the abstract logics $\mathbb{L}$ and $\mathbb{L}'$, then $\mathbb{L}/\ker h \cong \mathbb{L}'$ by means of a unique isomorphism g such that $h = g \circ \pi$, where π is the projection from $\mathbb{L}$ onto $\mathbb{L}/\ker h$.*

PROOF. See Brown and Suszko [1973] Theorem VIII.7.   ⊣

THEOREM 1.9 (Second Isomorphism Theorem). *If $\mathbb{L}$ is an abstract logic and θ, $\theta' \in \mathrm{Con}\,\mathbb{L}$ are such that $\theta \subseteq \theta'$, then $\theta'/\theta \in \mathrm{Con}\,(\mathbb{L}/\theta)$ and $(\mathbb{L}/\theta)/(\theta'/\theta) \cong \mathbb{L}/\theta'$, where the isomorphism is given in the customary way by $(a/\theta)/(\theta'/\theta) \mapsto a/\theta'$.*

PROOF. From the Second Isomorphism Theorem of Universal Algebra we know that the mapping $h : (\boldsymbol{A}/\theta)/(\theta'/\theta) \to \boldsymbol{A}/\theta'$ given by $h(a/\theta)/(\theta'/\theta) = a/\theta'$ is an isomorphism between the two algebras such that the following diagram commutes,

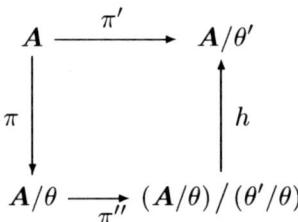

where π, π' and π'' are the natural projections. By construction we know that π and π' are also bilogical morphisms between the corresponding abstract logics. On the other hand, one can check that $\theta'/\theta \in \mathrm{Con}\,(\mathbb{L}/\theta)$, using that $\theta, \theta' \in \mathrm{Con}\,\mathbb{L}$. Thus π'' is also a bilogical morphism between $\mathbb{L}/\theta$ and $(\mathbb{L}/\theta)/(\theta'/\theta)$. Using all this, it is straightforward to check that

$$(\mathcal{C}/\theta)/(\theta'/\theta) = \left\{ h^{-1}[S] : S \in \mathcal{C}/\theta' \right\}$$

and as a consequence h, which we know is an algebraic isomorphism, is at the same time a bilogical morphism, that is, h is a logical isomorphism.   ⊣

THEOREM 1.10 (Correspondence Theorem). *If $\mathbb{L}$ is an abstract logic then for any $\theta \in \mathrm{Con}\,\mathbb{L}$, the segment $\left[\,\theta\,,\,\widetilde{\Omega}(\mathbb{L})\,\right]$ of $\mathrm{Con}\,\mathbb{L}$ is isomorphic to the lattice $\mathrm{Con}(\mathbb{L}/\theta)$ by the mapping $\theta' \mapsto \theta'/\theta$.*

PROOF. By Theorem 1.9 we know that if $\theta \subseteq \theta' \subseteq \widetilde{\Omega}(\mathbb{L})$, then $\theta'/\theta \in \mathrm{Con}(\mathbb{L}/\theta)$. Taking into account the Correspondence Theorem of Universal Algebra, it suffices to prove that, if $\theta \subseteq \theta' \in \mathrm{Con}\,A$ and $\theta'/\theta \in \mathrm{Con}(\mathbb{L}/\theta)$ then $\theta' \in \mathrm{Con}\,\mathbb{L}$: If $\langle a,b\rangle \in \theta'$ then $\langle a/\theta, b/\theta\rangle \in \theta'/\theta$ and therefore $C/\theta(a/\theta) = C/\theta(b/\theta)$, but since the projection is a bilogical morphism between $\mathbb{L}$ and $\mathbb{L}/\theta$ this implies $C(a) = C(b)$. This holds for any $a, b \in A$, so $\theta' \in \mathrm{Con}\,\mathbb{L}$.   ⊣

COROLLARY 1.11. *Let $\mathbb{L}$ be an abstract logic, and let $\theta \in \mathrm{Con}\,\mathbb{L}$. Then $\widetilde{\Omega}(\mathbb{L}/\theta) = \widetilde{\Omega}(\mathbb{L})/\theta$.*   ⊣

Thus for any $\mathbb{L}$, $\widetilde{\Omega}(\mathbb{L}/\widetilde{\Omega}(\mathbb{L})) = \widetilde{\Omega}(\mathbb{L})/\widetilde{\Omega}(\mathbb{L})$ is the identity on $A/\widetilde{\Omega}(\mathbb{L})$. This makes the following definition inevitable and natural:

DEFINITION 1.12. *An abstract logic $\mathbb{L} = \langle A, C\rangle$ is **reduced** when it has only one logical congruence, that is, when $\widetilde{\Omega}(\mathbb{L}) = Id_A$.*

*Given any abstract logic $\mathbb{L}$, we will put $\mathbf{L}^* = \mathbb{L}/\widetilde{\Omega}(\mathbb{L})$, and will call the abstract logic $\mathbf{L}^*$ **the reduction** of $\mathbb{L}$.*

If $\mathbf{L}$ is a class of abstract logics, then we will also put $\mathbf{L}^* = \{\,\mathbf{L}^* : \mathbb{L} \in \mathbf{L}\,\}$.

If $\mathbb{L}$ is an abstract logic, then Corollary 1.11 tells us that $\mathbf{L}^*$ is always reduced. If $\mathbb{L}$ is already reduced, then it is trivially isomorphic to its reduction $\mathbf{L}^*$, and one can simply identify both abstract logics. Given an abstract logic $\mathbb{L} = \langle A, C\rangle$, if we do not consider any other abstract logic on A, then there is no possible confusion if we write $A^* = A/\widetilde{\Omega}(\mathbb{L})$ with universe $A^* = A/\widetilde{\Omega}(\mathbb{L})$, and also $C^* = C/\widetilde{\Omega}(\mathbb{L})$ and $\mathcal{C}^* = \mathcal{C}/\widetilde{\Omega}(\mathbb{L})$, in order to write $\mathbf{L}^* = \langle A^*, C^*\rangle$; the projection will be $\pi(a) = a^* = a/\widetilde{\Omega}(\mathbb{L})$. These notational conventions will be used extensively throughout the monograph. The most elementary properties of this process of reduction follow (but see also Theorems 2.36 and 2.44):

PROPOSITION 1.13. *Assume that $\mathbb{L}$ is an abstract logic, and that $\theta \in \mathrm{Con}\,\mathbb{L}$. Then $(\mathbb{L}/\theta)^* \cong \mathbf{L}^*$.*

PROOF. Just consider the chain of equalities

$$(\mathbb{L}/\theta)^* = (\mathbb{L}/\theta)\,/\,\widetilde{\Omega}(\mathbb{L}/\theta) = (\mathbb{L}/\theta)\,/\,(\widetilde{\Omega}(\mathbb{L})/\theta) \cong \mathbb{L}/\widetilde{\Omega}(\mathbb{L}) = \mathbf{L}^*,$$

where we have used Corollary 1.11 and Theorem 1.9.   ⊣

PROPOSITION 1.14. *If there is a bilogical morphism between two abstract logics $\mathbb{L}$ and $\mathbb{L}'$ then their reductions are isomorphic, that is, $\mathbf{L}^* \cong (\mathbf{L}')^*$.*

PROOF. Let h be any bilogical morphism between $\mathbb{L}$ and $\mathbb{L}'$. By Theorem 1.8 we know that $\mathbb{L}/\ker h \cong \mathbb{L}'$. Since by Proposition 1.7 any logical isomorphism between two abstract logics puts their Tarski congruences into correspondence, their reductions are also isomorphic. That is, $(\mathbb{L}/\ker h)^* \cong (\mathbb{L}')^*$. Moreover we know that $\ker h \in \mathrm{Con}\,\mathbb{L}$, therefore we can apply Proposition 1.13 to obtain $(\mathbb{L}/\ker h)^* \cong \mathbb{L}^*$, and as a consequence $\mathbb{L}^* \cong (\mathbb{L}')^*$. $\dashv$

Therefore, the only possible bilogical morphisms between two reduced abstract logics are logical isomorphisms. The following result, which we will use in Chapter 2, is analogous to Theorem VIII.5 of Brown and Suszko [1973].

PROPOSITION 1.15. *Let* $\mathbb{L}$, $\mathbb{L}'$ *and* $\mathbb{L}''$ *be abstract logics, let* f *be a logical morphism from* $\mathbb{L}$ *to* $\mathbb{L}'$ *and let* g *be a bilogical morphism from* $\mathbb{L}$ *onto* $\mathbb{L}''$ *such that* $\ker g \subseteq \ker f$. *Then there is a unique logical morphism* h *from* $\mathbb{L}''$ *into* $\mathbb{L}'$ *such that* $h \circ g = f$. *Moreover,* f *projectively generates* $\mathbb{L}$ *from* $\mathbb{L}'$ *if and only if* h *projectively generates* $\mathbb{L}''$ *from* $\mathbb{L}'$.

PROOF. Let h be the unique homomorphism from $\boldsymbol{A}''$ into $\boldsymbol{A}'$ such that $h \circ g = f$; its existence is guaranteed by the condition that $\ker g \subseteq \ker f$. If $T \in \mathcal{C}'$ then $g^{-1}\big[h^{-1}[T]\big] = f^{-1}[T] \in \mathcal{C}$ since f is a logical morphism; but since g is a bilogical morphism, this implies that $h^{-1}[T] \in \mathcal{C}''$, and thus we see that h is also a logical morphism. If, moreover, f projectively generates $\mathbb{L}$ from $\mathbb{L}'$, then Theorem VIII.5 of Brown and Suszko [1973] proves that h also projectively generates $\mathbb{L}''$ from $\mathbb{L}'$. If, conversely, h projectively generates $\mathbb{L}''$ from $\mathbb{L}'$, then using that g is a bilogical morphism, we have $\mathcal{C} = \{g^{-1}[S] : S \in \mathcal{C}''\} = \{g^{-1}\big[h^{-1}[T]\big] : T \in \mathcal{C}'\}$, that is, $\mathcal{C} = \{f^{-1}[T] : T \in \mathcal{C}'\}$ which says that f projectively generates $\mathbb{L}$ from $\mathbb{L}'$. $\dashv$

It is a well-known result of Universal Algebra that any algebra is isomorphic to a quotient of a formula algebra constructed from a large enough set of variables. This fact extends to abstract logics in the following sense:

PROPOSITION 1.16. *Let* $\mathbb{L}$ *be an abstract logic and* κ *an infinite cardinal number,* $\kappa \geqslant \mathrm{card}\, A_{\mathbb{L}}$. *If we denote by* $\boldsymbol{Fm}_\kappa$ *the algebra of formulas with* κ *variables, then there is an abstract logic* $\mathbb{L}_\kappa$ *on* $\boldsymbol{Fm}_\kappa$ *and a congruence* $\theta \in \mathrm{Con}\,\mathbb{L}_\kappa$ *such that* $\mathbb{L}$ *is isomorphic to* $\mathbb{L}_\kappa/\theta$.

PROOF. Let $h : \boldsymbol{Fm}_\kappa \to \boldsymbol{A}_{\mathbb{L}}$ be any epimorphism. We can consider the abstract logic $\mathbb{L}_\kappa$ projectively generated from $\mathbb{L}$ by h; then h is a bilogical morphism from $\mathbb{L}_\kappa$ onto $\mathbb{L}$, and by Proposition 1.4 $\ker h \in \mathrm{Con}\,\mathbb{L}_\kappa$; therefore the Homomorphism Theorem 1.8 tells us that $\mathbb{L}_\kappa/\ker h \cong \mathbb{L}$. $\dashv$

Since it seems clear that logical morphisms are one of the right kind of "morphisms" between abstract logics, and that bilogical morphisms determine some kind of "equivalence" between abstract logics, it is important to determine which properties are preserved under bilogical morphisms. It turns out that many typical metalogical properties of closure operators, like the Deduction Theorem or the Property of Disjunction, satisfy this requirement, see Section 2.4. We prove here a very basic one:

PROPOSITION 1.17. *If h is a bilogical morphism between two abstract logics* $\mathbb{L}$ *and* $\mathbb{L}'$, *then one of them is finitary if and only if the other one is finitary.*

PROOF. Just use the following two facts, already established in Proposition 1.4: that $C(X) = h^{-1}[C'(h[X])]$ and that $C'(X) = h[C(h^{-1}[X])]$. $\quad\dashv$

Note that this property cannot be proved by using Proposition 1.5 alone, because the lattice isomorphism induced by h between the corresponding closure systems might not preserve unions of directed families; thus the proof published in Verdú [1987] is erroneous. Indeed, while it is true that if $\mathcal{C}$ is an inductive closure system then the lattice $\langle \mathcal{C}, \subseteq \rangle$ is algebraic, it is interesting to note that the converse might not be true: if for some closure system $\mathcal{C}$ the lattice $\langle \mathcal{C}, \subseteq \rangle$ is algebraic, then it is isomorphic to the closure system of closed sets of some finitary closure operator, but this operator might not be the original one; this fact has been recognized recently by Herrmann in the context of his generalization of Blok and Pigozzi's theory of algebraizable logics to non-finitary ones, see Herrmann [1993b] and Herrmann and Wolter [1994].

Sentential logics

It is customary to define a ***sentential logic*** as a pair of the form $\mathcal{S} = \langle \boldsymbol{Fm}, \vdash_{\mathcal{S}} \rangle$ where $\boldsymbol{Fm}$ is a formula algebra, and $\vdash_{\mathcal{S}} \subseteq P(Fm) \times Fm$ is a relation satisfying the following five properties, for all $\Gamma, \Delta \subseteq Fm$ and all $\varphi \in Fm$ (as usual we write $\Gamma \vdash_{\mathcal{S}} \varphi$ for $\langle \Gamma, \varphi \rangle \in \vdash_{\mathcal{S}}$):

(S1) If $\varphi \in \Gamma$ then $\Gamma \vdash_{\mathcal{S}} \varphi$.
(S2) If $\Gamma \vdash_{\mathcal{S}} \varphi$ and $\Gamma \subseteq \Delta$ then $\Delta \vdash_{\mathcal{S}} \varphi$.
(S3) If $\Gamma \vdash_{\mathcal{S}} \varphi$ and for every $\gamma \in \Gamma$, $\Delta \vdash_{\mathcal{S}} \gamma$ then $\Delta \vdash_{\mathcal{S}} \varphi$.
(S4) If $\Gamma \vdash_{\mathcal{S}} \varphi$ then there is a finite $\Gamma_0 \subseteq \Gamma$ such that $\Gamma_0 \vdash_{\mathcal{S}} \varphi$.
(S5) If $\Gamma \vdash_{\mathcal{S}} \varphi$ then $e[\Gamma] \vdash_{\mathcal{S}} e(\varphi)$ for all substitutions $e \in \text{Hom}(\boldsymbol{Fm}, \boldsymbol{Fm})$.

Note that property (S2) is a consequence of properties (S1) and (S3). In general, a relation satisfying properties (S1) to (S3) is called a ***consequence relation***, while property (S4) is called *finitarity*, and condition (S5) is called *structurality*[12]; thus in this monograph we define a sentential logic as a *finitary and structural consequence relation on a formula algebra*[13].

The notation $\Gamma \vdash_S \Delta$ means that $\Gamma \vdash_S \delta$ holds for all $\delta \in \Delta$; remark that this notation has nothing to do with "multiple-conclusion" consequences. The notation $\Gamma \dashv\vdash_S \Delta$ means that both $\Gamma \vdash_S \Delta$ and $\Delta \vdash_S \Gamma$ hold.

In order to present sentential logics as a particular kind of abstract logics, we can equally say that a sentential logic is an abstract logic $\mathcal{S} = \langle \boldsymbol{Fm}, \mathrm{Cn}_S \rangle$ on an algebra of formulas such that the closure operator Cn_S is finitary and structural; for a closure operator to be **structural** means that, for every $e \in \mathrm{Hom}(\boldsymbol{Fm}, \boldsymbol{Fm})$, if $\varphi \in \mathrm{Cn}_S(\Gamma)$ then also $e(\varphi) \in \mathrm{Cn}_S(e[\Gamma])$. The equivalence between the two definitions is easily established by setting:

$$\varphi \in \mathrm{Cn}_S(\Gamma) \iff \Gamma \vdash_S \varphi \tag{1.4}$$

The closed sets of the operator Cn_S are called the **theories** of the sentential logic, and the closure system they form is denoted by $Th\mathcal{S}$; the property of being structural can be formulated in terms of theories by saying that the family $Th\mathcal{S}$ is closed under inverse substitutions, i.e., if $\Gamma \in Th\mathcal{S}$ then $e^{-1}[\Gamma] \in Th\mathcal{S}$ for any substitution e. In informal remarks we often refer to a sentential logic as a **logical system** or simply as a **logic**.

Since we treat a sentential logic as a special kind of abstract logic, all previous notions and results concerning finitary abstract logics apply to them; but in addition a sentential logic is also structural. This implies that the set of theorems $\mathrm{Cn}_S(\emptyset)$ is closed under substitutions. However, note that an arbitrary theory need not be so; therefore, whenever we consider axiomatic extensions $\mathcal{S}^\Gamma$ of $\mathcal{S}$ by some $\Gamma \in Th\mathcal{S}$ in the sense defined on page 18, we are referring to the abstract logic $\langle \boldsymbol{Fm}, \mathcal{S}^\Gamma \rangle$, but this one will be structural (i.e., a sentential logic) if and only if Γ is closed under substitutions; this is for instance the case whenever Γ is the theory generated by a set of additional axioms closed under substitutions (sometimes called *axiom schemes*), which is the most common situation.

---

[12]Condition (S5) is also called, equally often, *substitution invariance*.

[13]In other, more comprehensive studies in the general theory of abstract algebraic logic (such as Czelakowski [2001b]) the property of finitarity is not incorporated into the definition of a sentential logic, but is rather one of its possible properties subject to investigation.

S-filters and S-matrices

Given a sentential logic S and an algebra A of the same similarity type, a subset $F \subseteq A$ is an S-*filter*[14] iff for any $\Gamma \cup \{\varphi\} \subseteq Fm$ and any interpretation $h \in \mathrm{Hom}(Fm, A)$,

$$\text{if } \Gamma \vdash_S \varphi \text{ and } h[\Gamma] \subseteq F \text{ then } h(\varphi) \in F.$$

Observe that another, often practical way of saying the same thing is that for any $h \in \mathrm{Hom}(Fm, A)$, the set $h^{-1}[F]$ is a theory of S. A matrix $\langle A, F \rangle$ is a *matrix for* S, or an S-matrix, when F is an S-filter on A; the class of all S-matrices will be denoted by **Matr**S, and the class of all reduced S-matrices by **Matr**$^*S$. The set of all S-filters on a given algebra A is denoted by $\mathcal{F}i_S A$; this set is an inductive closure system, thus it is also an algebraic, and hence complete, lattice, ordered by $\subseteq$. The associated closure operator will be denoted by Fi_S^A; that is, for any $X \subseteq A$, $\mathrm{Fi}_S^A(X)$ is the least subset of A containing X which is "closed under the inferences of S" in the sense that it is closed under the images of these inferences under any interpretation; more precisely, one has the following characterization, which will be useful at several points in the monograph:

LEMMA 1.18. *For all* $X \subseteq A$, $\mathrm{Fi}_S^A(X) = \bigcup \{X_n : n \in \omega\}$ *where the sets* X_n *are defined as follows:* $X_0 = X$, *and for any* $n \in \omega$, $X_{n+1} = \{x \in A : \text{There are } \varphi \in Fm \text{ and a finite } \Gamma \subseteq Fm \text{ such that } \Gamma \vdash_S \varphi \text{ and there is } h \in \mathrm{Hom}(Fm, A) \text{ with } h[\Gamma] \subseteq X_n \text{ and } h(\varphi) = x\}$. $\dashv$

The following facts will be used later on:

PROPOSITION 1.19. *Let* $h : A \to B$ *be an (algebraic) homomorphism. Then, for any* S-*filter* G *on* B, $h^{-1}[G]$ *is an* S-*filter on* A; *and if moreover* h *is surjective then for any* $G \subseteq B$, *if* $h^{-1}[G]$ *is an* S-*filter on* A *then also* G *is an* S-*filter on* B.

PROOF. If G is an S-filter on B, taking the comment that follows the definition of S-filter into consideration it is easy to see that $h^{-1}[G]$ is an S-filter on A. Now, if $G \subseteq B$ is such that $h^{-1}[G]$ is an S-filter, and $\Gamma \vdash_S \varphi$, let $g \in \mathrm{Hom}(Fm, B)$ be such that $g[\Gamma] \subseteq G$. By the Axiom of Choice, there is $f \in \mathrm{Hom}(Fm, A)$ such that $h \circ f = g$. Therefore, $f[\Gamma] \subseteq h^{-1}[G]$; so $f(\varphi) \in h^{-1}[G]$ and hence, $g(\varphi) \in G$. This proves that G is an S-filter on B. $\dashv$

---

[14] In some cases, especially where the notion of S-filter coexists with a purely algebraic notion of filter (such as lattice filters in any kind of algebras having a lattice reduct), the terms *logical filter* and *deductive filter* are also used for emphasis; the latter originates in Rasiowa [1974], p. 200.

PROPOSITION 1.20. *If $F \in \mathcal{F}i_S A$ and $\theta \in \mathrm{Con}\, A$, then θ is compatible with F, that is, $\theta \subseteq \Omega_A(F)$, if and only if $F = \pi_\theta^{-1}[G]$ for some $G \in \mathcal{F}i_S(A/\theta)$, where π_θ is the projection from A onto A/θ.*

PROOF. If $\theta \in \mathrm{Con}\, A$ is compatible with $F \in \mathcal{F}i_S A$, then $\pi_\theta^{-1}\big[\pi_\theta[F]\big] = F$. Therefore, by Proposition 1.19, $G = \pi_\theta[F]$ is the required S-filter. Conversely, if $F = \pi_\theta^{-1}[G]$ for some $G \in \mathcal{F}i_S A$, $\langle a, b \rangle \in \theta$ and $a \in F$, then $\pi_\theta(b) = \pi_\theta(a) \in G$. So, $b \in F$, and thus we see that θ is compatible with F.   ⊣

PROPOSITION 1.21. *Let $h : A \to B$ be an epimorphism. Then the following conditions are equivalent:*

(i) *h is a bilogical morphism between the abstract logic $\langle A, \mathcal{F}i_S A \rangle$ and the abstract logic $\langle B, \mathcal{F}i_S B \rangle$.*

(ii) *h induces an isomorphism between the lattices $\mathcal{F}i_S A$ and $\mathcal{F}i_S B$.*

(iii) *For any $F \in \mathcal{F}i_S A$, $h[F] \in \mathcal{F}i_S B$, and $\ker h \in \mathrm{Con}\,\langle A, \mathcal{F}i_S A \rangle$.*

PROOF. Clearly (i) implies (ii), and also (ii) implies (iii) since if $a, b \in A$ satisfy $h(a) = h(b)$ and $a \in F \in \mathcal{F}i_S A$ then $b \in h^{-1}\big[h[F]\big] = F$. Now suppose that (iii) holds; then by assumption $h[F] \in \mathcal{F}i_S B$ for all $F \in \mathcal{F}i_S A$, and conversely for every $G \in \mathcal{F}i_S B$ we know that $h^{-1}[G] \in \mathcal{F}i_S A$ and $G = h\big[h^{-1}[G]\big]$ because h is surjective. Then 1.4.(v) shows that h is a bilogical morphism.   ⊣

PROPOSITION 1.22. *If h is a bilogical morphism from $\langle A, \mathcal{F}i_S A \rangle$ onto $\langle B, C \rangle$ then $C = \mathcal{F}i_S B$. In particular, $(\mathcal{F}i_S A)^* = \mathcal{F}i_S(A^*)$. Moreover, if two abstract logics $\mathbb{L} = \langle A, C \rangle$ and $\mathbb{L}' = \langle A', C' \rangle$ are isomorphic, then $C = \mathcal{F}i_S A$ if and only if $C' = \mathcal{F}i_S A'$.*

PROOF. By Proposition 1.19 $C \subseteq \mathcal{F}i_S B$; on the other hand if $F \in \mathcal{F}i_S B$ it is always true that $h^{-1}[F] \in \mathcal{F}i_S A$, but since h is a bilogical morphism, $F = h\big[h^{-1}[F]\big] \in C$. The last part can be proved by applying the first one both to h and to h^{-1}.   ⊣

The classes Alg*S and K$_S$

If $\mathcal{M} = \langle A, D \rangle \in \mathbf{Matr}S$ and $\theta \in \mathrm{Con}\, A$ is compatible with D, then also $\mathcal{M}/\theta = \langle A/\theta, D/\theta \rangle \in \mathbf{Matr}S$, because $\pi_\theta^{-1}[D/\theta] = D$; in particular $\mathcal{M}^* \in \mathbf{Matr}^*S$ and it is then easy to show that $\mathbf{Matr}^*S$ is the closure under isomorphisms of the class $(\mathbf{Matr}S)^*$. We will denote by $\mathbf{Alg}^*S$ the class of **all algebra reducts of all matrices in** $\mathbf{Matr}^*S$. This is the class of algebras usually associated with any sentential logic; for instance if S is algebraizable in the sense of Blok and Pigozzi [1989a], then $\mathbf{Alg}^*S$ is the equivalent quasivariety semantics of S. In Rasiowa [1974], for the systems there considered (which are all alge-braizable), the algebras in $\mathbf{Alg}^*S$ are called "S-algebras"; in Chapter 2 we will extend this term to cover more cases; see also the comments on page 36 after Definition 2.16. Note specially that the class $\mathbf{Alg}^*S$ is *not* the result of applying the reduction process to any other class of algebras: in general we apply the star notation to classes of matrices and of abstract logics to indicate the result of the reduction process, but the class of "S-algebras" that will be introduced in Section 2.2, and which will be denoted by $\mathbf{Alg}S$, will bear a different relation to $\mathbf{Alg}^*S$; however, in choosing the notation $\mathbf{Alg}^*S$ we have preferred to follow the standard practice.

Note that for any $F \in \mathcal{F}i_S A$, $\Omega_A(F) \in \mathrm{Con}_{\mathbf{Alg}^*S} A$. In Blok and Pigozzi [1992] the authors characterize several kinds of sentential logics[15] by the be-haviour of the Leibniz operator Ω_A with respect to the lattice structures of $\mathcal{F}i_S A$ and of $\mathrm{Con}_{\mathbf{Alg}^*S} A$ for an arbitrary algebra A, a trend already advanced in Blok and Pigozzi [1986], [1989a]. Some of their results will be used in this monograph.

Due to the fact that S is structural, the S-filters on the formula algebra are exactly the S-theories; and the characterization (1.1) of the Leibniz congruence on page 16 takes the following simpler form on $\boldsymbol{Fm}$, already found in Łoś [1949]: If $\Gamma \subseteq Fm$ then for every $\varphi, \psi \in Fm$,

$$\langle \varphi, \psi \rangle \in \Omega_{\boldsymbol{Fm}}(\Gamma) \iff \forall \gamma(p, \vec{q}) \in Fm,$$
$$\gamma(\varphi, \vec{q}) \in \Gamma \text{ iff } \gamma(\psi, \vec{q}) \in \Gamma. \tag{1.5}$$

As a consequence, the characterization (1.3) of the Tarski congruence on page 19, becomes in the case of a sentential logic

$$\langle \varphi, \psi \rangle \in \widetilde{\Omega}(S) \iff \forall \gamma(p, \vec{q}) \in Fm, \ \gamma(\varphi, \vec{q}) \dashv\vdash_S \gamma(\psi, \vec{q}) \tag{1.6}$$

(this characterization appears already in Smiley [1962] and in Wójcicki [1988] p. 59, although with different terminology and notation).

As we have already commented on page 19, in the case of a sentential logic S the Tarski congruence $\widetilde{\Omega}(S)$ is actually the one normally used to obtain the so-called ***Lindenbaum-Tarski algebra*** of S, which is $\boldsymbol{Fm}^* = \boldsymbol{Fm}/\widetilde{\Omega}(S)$; ac-cordingly, one can call the abstract logic $S^* = \langle \boldsymbol{Fm}^*, \mathrm{Cn}_S/\widetilde{\Omega}(S) \rangle$ the ***Lin-denbaum-Tarski quotient*** of S. We will denote by $\mathbf{K}_S$ ***the variety generated by the Lindenbaum-Tarski algebra*** $\boldsymbol{Fm}^*$. This variety is sometimes considered to be the class of algebras canonically associated with S, as in Rautenberg [1991]. However, there are examples in the literature where the class $\mathbf{Alg}^*S$, associated with S in the general theory of matrices, is not a variety but a quasivariety, or

---

[15] The classes of logics that result and the relations between them form what has been called later on ***the Leibniz hierarchy***; see Font, Jansana, and Pigozzi [2003], page 49, for a picture.

even a non-elementary class. It is well-known (see Wójcicki [1988] Lemma 1.7.4) that $\widetilde{\Omega}(\mathcal{S})$ is a fully invariant congruence of $\boldsymbol{Fm}$; as a consequence an equation $\varphi \approx \psi$ holds in $\mathbf{K}_{\mathcal{S}}$, that is, it holds in $\boldsymbol{Fm}^* = \boldsymbol{Fm}/\widetilde{\Omega}(\mathcal{S})$, iff $\langle \varphi, \psi \rangle \in \widetilde{\Omega}(\mathcal{S})$, and the algebra $\boldsymbol{Fm}^*$ is free in $\mathbf{K}_{\mathcal{S}}$ (see Burris and Sankappanavar [1981] Lemma 14.7 for instance).

Matrices are used to build up a *semantics* for sentential logics, and the usual completeness notion arises: one says that a sentential logic $\mathcal{S}$ is **complete with respect to** a class $\mathbf{M}$ of matrices when for all $\Gamma \cup \{\varphi\} \subseteq Fm$, $\Gamma \vdash_{\mathcal{S}} \varphi$ holds if and only if for every matrix $\langle \boldsymbol{A}, F \rangle \in \mathbf{M}$ and every $h \in \mathrm{Hom}(\boldsymbol{Fm}, \boldsymbol{A})$, $h[\Gamma] \subseteq F$ implies $h(\varphi) \in F$. From the fact that $\mathcal{F}i_{\mathcal{S}} \boldsymbol{Fm} = Th\mathcal{S}$ it immediately follows that an arbitary sentential logic $\mathcal{S}$ is complete with respect to the whole class $\mathbf{Matr}\mathcal{S}$; one can also prove that any $\mathcal{S}$ is complete with respect to the class $\mathbf{Matr}^*\mathcal{S}$. For these and related questions on matrix semantics see Wójcicki [1988]. We will just need the following result:

PROPOSITION 1.23. $\mathbf{K}_{\mathcal{S}}$ *is the variety generated by the class* $\mathbf{Alg}^*\mathcal{S}$.

PROOF. As we have noted, an equation $\varphi \approx \psi$ holds in $\mathbf{K}_{\mathcal{S}}$ iff $\langle \varphi, \psi \rangle \in \widetilde{\Omega}(\mathcal{S})$, that is, by (1.6), iff for any $\gamma(p, \vec{q}) \in Fm$, $\gamma(\varphi, \vec{q}) \dashv\vdash_{\mathcal{S}} \gamma(\psi, \vec{q})$. Since $\mathcal{S}$ is complete with respect to the class $\mathbf{Matr}^*\mathcal{S}$, this holds iff for any $\langle \boldsymbol{A}, F \rangle \in \mathbf{Matr}^*\mathcal{S}$ and any sequences $\vec{a}, \vec{c}$ in $\boldsymbol{A}$, $\gamma^{\boldsymbol{A}}(\varphi^{\boldsymbol{A}}(\vec{a}), \vec{c}) \in F \iff \gamma^{\boldsymbol{A}}(\psi^{\boldsymbol{A}}(\vec{a}), \vec{c}) \in F$, which by (1.1) amounts to saying that for all $\vec{a}$, $\langle \varphi^{\boldsymbol{A}}(\vec{a}), \psi^{\boldsymbol{A}}(\vec{a}) \rangle \in \Omega_{\boldsymbol{A}}(F)$, and this is equivalent to $\varphi^{\boldsymbol{A}}(\vec{a}) = \psi^{\boldsymbol{A}}(\vec{a})$ because the matrix is reduced. Finally, to say that this holds for all reduced matrices of $\mathcal{S}$ is equivalent to saying that the equation $\varphi \approx \psi$ holds in every $\boldsymbol{A} \in \mathbf{Alg}^*\mathcal{S}$.   ⊣

The reader may have noticed that the same proof actually shows that the class of all algebra reducts of any class $\mathbf{M}$ of reduced matrices such that $\mathcal{S}$ is complete with respect to $\mathbf{M}$ generates the same variety $\mathbf{K}_{\mathcal{S}}$. We will find better descriptions of this class of algebras, for some restricted cases, in Section 2.4, and also in Chapter 4.

CHAPTER 2

ABSTRACT LOGICS AS MODELS
OF SENTENTIAL LOGICS

In this chapter we consider abstract logics as models of sentential logics. Abstract logics are suitable for modelling the metalogical properties that sentential logics can have; in this they differ notably from matrices. Our purpose is to single out for any sentential logic a class of abstract logics that exhibit some crucial metalogical properties of it. This leads us to distinguish two types of models for a sentential logic, the models "tout court" and the full models. The latter will be suitable for our purpose of modelling metalogical properties, an issue that will be dealt with specifically in the last section of this chapter, and also in Chapter 4.

2.1. Models and full models

We begin by using an abstract logic to define a logic on the algebra of formulas by the ordinary semantic procedure; using it the notion of model will be introduced.

DEFINITION 2.1. *If $\mathbb{L} = \langle A, C \rangle$ is any abstract logic, the relation $\models_{\mathbb{L}}$ induced by $\mathbb{L}$ on the formula algebra is defined, for any $\Gamma \cup \{\varphi\} \subseteq Fm$, by:*

$$\Gamma \models_{\mathbb{L}} \varphi \iff \text{for any } h \in \text{Hom}(\boldsymbol{Fm}, \boldsymbol{A}), \ h(\varphi) \in C(h[\Gamma]).$$

If L is any class of abstract logics, then it induces on the formula algebra the relation $\models_{\mathsf{L}} = \bigcap \{\models_{\mathbb{L}} : \mathbb{L} \in \mathsf{L}\}$.

PROPOSITION 2.2. *The relations $\models_{\mathbb{L}}$ and $\models_{\mathsf{L}}$ defined on the formula algebra $\boldsymbol{Fm}$ are structural consequence relations on this algebra.*

PROOF. It is easy to see that $\models_{\mathbb{L}}$ is a consequence relation, that is, that the operator defined as $\varphi \in \text{Cn}_{\mathbb{L}}(\Gamma)$ iff $\Gamma \models_{\mathbb{L}} \varphi$ is a closure operator on $\boldsymbol{Fm}$. Actually, $\text{Cn}_{\mathbb{L}}$ is the abstract logic on $\boldsymbol{Fm}$ projectively generated from $\mathbb{L}$ by the family of all homomorphisms $\text{Hom}(\boldsymbol{Fm}, \boldsymbol{A})$. By Theorem XII.2 of Brown and

Suszko [1973], it is structural. Moreover, the meet of any family of structural closure operators is also a structural closure operator. ⊣

PROPOSITION 2.3. *If there is a bilogical morphism between the abstract logics* $\mathbb{L}$ *and* $\mathbb{L}'$ *then* $\models_{\mathbb{L}} = \models_{\mathbb{L}'}$; *in particular,* $\models_{\mathbb{L}} = \models_{\mathbb{L}^*}$.

PROOF. Let $f : \boldsymbol{A} \to \boldsymbol{A}'$ be the epimorphism which is a bilogical morphism between $\mathbb{L}$ and $\mathbb{L}'$. Since for any $h \in \mathrm{Hom}(\boldsymbol{Fm}, \boldsymbol{A})$, $f \circ h \in \mathrm{Hom}(\boldsymbol{Fm}, \boldsymbol{A}')$, using 1.4.(ii) we get that $\models_{\mathbb{L}'} \subseteq \models_{\mathbb{L}}$. Conversely, given any $g \in \mathrm{Hom}(\boldsymbol{Fm}, \boldsymbol{A}')$, since f is surjective, there is (by the Axiom of Choice) an $h \in \mathrm{Hom}(\boldsymbol{Fm}, \boldsymbol{A})$ such that $f \circ h = g$; then if $\Gamma \models_{\mathbb{L}} \varphi$ we have $h(\varphi) \in \mathrm{C}(h[\Gamma])$ which implies $f(h(\varphi)) \in f[\mathrm{C}(h[\Gamma])]$ but since $f \circ \mathrm{C} = \mathrm{C}' \circ f$ by 1.4(iii), we obtain $g(\varphi) \in \mathrm{C}'(g[\Gamma])$. This proves $\models_{\mathbb{L}} \subseteq \models_{\mathbb{L}'}$. ⊣

Now we introduce the general notion of an abstract logic being a model of a sentential logic.

DEFINITION 2.4. *An abstract logic* $\mathbb{L}$ *is a **model** of a sentential logic* $\mathcal{S}$ *when for any* $\Gamma \cup \{\varphi\} \subseteq Fm$, $\Gamma \vdash_{\mathcal{S}} \varphi$ *implies* $\Gamma \models_{\mathbb{L}} \varphi$. *The class of all models of* $\mathcal{S}$ *will be denoted by* $\mathrm{Mod}\mathcal{S}$.

A sentential logic $\mathcal{S}$ *is **complete** with respect to a class of abstract logics* $\mathbf{L}$ *when for any* $\Gamma \cup \{\varphi\} \subseteq Fm$, $\Gamma \vdash_{\mathcal{S}} \varphi$ *iff* $\Gamma \models_{\mathbf{L}} \varphi$.

From Proposition 2.3 follows at once:

PROPOSITION 2.5.

(1) *If there is a bilogical morphism between* $\mathbb{L}$ *and* $\mathbb{L}'$ *then* $\mathbb{L}$ *is a model of* $\mathcal{S}$ *iff* $\mathbb{L}'$ *is; in particular,* $\mathbb{L}$ *is a model of* $\mathcal{S}$ *iff* $\mathbb{L}^*$ *is.*
(2) *If* $\mathcal{S}$ *is complete with respect to a class* $\mathbf{L}$ *of abstract logics, then it is also complete with respect to the class* $\mathbf{L}^*$. ⊣

The structurality of a sentential logic $\mathcal{S}$ implies that $\mathcal{S}$ is a model of itself, therefore so is its Lindenbaum-Tarski quotient $\mathcal{S}^* = \mathcal{S}/\widetilde{\mathit{\Omega}}(\mathcal{S})$; thus we have:

PROPOSITION 2.6. *A sentential logic* $\mathcal{S}$ *is complete with respect to any class* $\mathbf{L}$ *of its models that includes either* $\mathcal{S}$ *or* $\mathcal{S}^*$, *and also with respect to the corresponding reduced class* $\mathbf{L}^*$. *In particular,* $\mathcal{S}$ *is complete with respect to the class of all its models, and also with respect to the class of all its reduced models.* ⊣

Since $h(\varphi) \in \mathrm{C}(h[\Gamma])$ if and only if $h(\varphi) \in T$ for every $T \in \mathcal{C}$ such that $h[\Gamma] \subseteq T$, it results at once that:

PROPOSITION 2.7. *An abstract logic* $\mathbb{L} = \langle A, \mathcal{C} \rangle$ *is a model of a sentential logic* S *if and only if for every* $T \in \mathcal{C}$, *the matrix* $\langle A, T \rangle$ *is a matrix for* S; *that is, if and only if* $\mathcal{C} \subseteq \mathcal{F}i_S A$. $\dashv$

Thus for every algebra A, the whole family $\mathcal{F}i_S A$ determines a model of S on A having the finest closure system; therefore this model is the *weakest* model of S on A, according to the ordering relation between abstract logics defined on page 18.

The notion of model we have just defined corresponds to the notion of a *generalized matrix of* a sentential logic, defined by Wójcicki as an arbitrary family of matrices over the same algebra in his [1969], [1973]. It is obvious that such a family is a generalized matrix for some S if and only if the abstract logic whose closure system is the one generated by the set of filters of the matrices in that family is a model of S in our sense, and conversely every model of S can be thought of as a generalized matrix for S. The same notion of model, in the form of a closure operator rather than of a closure system, was put forward by Smiley in [1962].

In principle it might seem that this notion of model is finer than the usual one (a matrix), since each model possesses the same structure (a closure operator) which the sentential logic has; actually with its help one can express the notion of "being a model of a Gentzen-style rule" in a direct way (see Definition 4.3). However, we have seen that *any* family of matrices makes a model; due to this fact, models can be wildly different from what we intend them to model, and they might not exhibit some crucial metalogical properties of a sentential logic, like the Property of Disjunction or the Deduction Theorem, as discussed in Section 2.4 and in Chapter 4. For this reason we will define a more restricted kind of models.

DEFINITION 2.8. *If* S *is a sentential logic, then an abstract logic* $\mathbb{L} = \langle A, \mathcal{C} \rangle$ *is a **full model of** S iff* $\mathbb{L}^*$ *is equal to the abstract logic* $\langle A^*, \mathcal{F}i_S A^* \rangle$, *that is, iff the closure system of the reduction of* $\mathbb{L}$ *consists of all the* S-*filters of the quotient algebra.*

The class of all full models of S *will be denoted by* **FMod**S, *and the class of all reduced full models of* S *by* **FMod**$^*S$; *and for each algebra* A, *the set of all full models of* S *on* A *will be denoted by* $\mathcal{F}Mod_S A$.

We begin our study of full models by confirming that they are indeed models of the sentential logic, thus justifying the use of the term *model* in the name we have chosen for this notion. Moreover, we see that they inherit some properties of the sentential logic they model:

PROPOSITION 2.9. *Let* $\mathbb{L}$ *be a full model of a sentential logic* S. *Then:*

(1) $\mathbb{L}$ *is a model of* S.
(2) $\mathbb{L}$ *is finitary.*
(3) $\mathbb{L}$ *has theorems if and only if* S *has theorems.*

PROOF. If $\mathbb{L} = \langle A, C \rangle$ is a full model of S, then $\mathbb{L}^* = \langle A^*, \mathcal{F}i_S A^* \rangle$; but an abstract logic of this kind is always finitary (because the union of an upwards directed family of S-filters is an S-filter), and by 2.7 it is a model of S; since the canonical projection from A onto A^* is a bilogical morphism, by 1.17 and 2.5, $\mathbb{L}$ itself will also be finitary and a model of S, that is, (1) and (2) hold. If S does not have theorems then the empty set is the least S-filter on any algebra, and thus it must be a closed set of any full model of S. Conversely, if S has theorems then any S-filter has to be non-empty, in particular the least closed set of any full model of S. This proves (3).   ⊣

It is not true that every model is a full model: see Section 5.1.1. Actually, an interesting problem is to find necessary and/or sufficient conditions for a model to be full which are at the same time logically interesting and useful for applications. In Sections 4.2 and 4.3 we solve this problem for two particular classes of sentential logics. Let us continue with elementary properties of full models of a sentential logic.

PROPOSITION 2.10. *For any algebra* A, *the abstract logic* $\langle A, \mathcal{F}i_S A \rangle$ *is a full model of* S, *and it is indeed the weakest full model of* S *on* A *(i.e., the one having the finest closure system)*[16].

PROOF. If we consider the reduction $\langle A^*, (\mathcal{F}i_S A)^* \rangle$ of $\langle A, \mathcal{F}i_S A \rangle$, then the canonical projection π is a bilogical morphism, so by 1.22 $(\mathcal{F}i_S A)^* = \mathcal{F}i_S A^*$. As a consequence, $\langle A, \mathcal{F}i_S A \rangle$ is a full model of S. And by 2.9 it is obviously the weakest one since it is simply the weakest model of S.   ⊣

In particular any sentential logic is a full model of itself, and it is actually the weakest one on Fm.

PROPOSITION 2.11. *The class* **FMod**S *is closed under bilogical morphisms: That is, if there is a bilogical morphism between two abstract logics* $\mathbb{L}_1$ *and* $\mathbb{L}_2$ *then* $\mathbb{L}_1$ *is a full model of* S *if and only if* $\mathbb{L}_2$ *is a full model of* S. *In particular, an abstract logic* $\mathbb{L}$ *is a full model of* S *if and only if its reduction* $\mathbb{L}^*$ *is.*

---

[16]The full models of the form $\langle A, \mathcal{F}i_S A \rangle$ for an arbitrary algebra A have been called **basic full models** of S in the later literature, beginning with Definition 2.10(i) in Font, Jansana, and Pigozzi [2001].

PROOF. If there is a bilogical morphism between $\mathbb{L}_1$ and $\mathbb{L}_2$ then $\mathbb{L}_1^*$ is (logically) isomorphic to $\mathbb{L}_2^*$. If one of them, say $\mathbb{L}_1$, is a full model of $\mathcal{S}$, then $\mathcal{C}_1^* = \mathcal{F}i_\mathcal{S} A_1^*$ and by 1.22 also $\mathcal{C}_2^* = \mathcal{F}i_\mathcal{S} A_2^*$; but since $\mathbb{L}_2^*$ is reduced, this implies that $\mathbb{L}_2$ is a full model of $\mathcal{S}$.   ⊣

From Definition 2.8 and Propositions 2.10 and 2.11 it results at once:

COROLLARY 2.12. *An abstract logic $\mathbb{L}$ is a full model of $\mathcal{S}$ if and only if there is a bilogical morphism from $\mathbb{L}$ onto an abstract logic of the form $\langle B, \mathcal{F}i_\mathcal{S} B \rangle$.*   ⊣

COROLLARY 2.13. *The class $\mathbf{FMod}\mathcal{S}$ is the smallest class of abstract logics that contains all those of the form $\langle B, \mathcal{F}i_\mathcal{S} B \rangle$ and is closed under bilogical morphisms (i.e., under the operations of taking images and inverse images by bilogical morphisms).*   ⊣

We will use these facts very often, namely when we want to prove that some property holds for every full model: If the property is preserved under bilogical morphisms, then it is enough, and often simpler, to prove that it holds for every abstract logic of the form $\langle B, \mathcal{F}i_\mathcal{S} B \rangle$. Each of these abstract logics is the finest full model on the corresponding algebra, and Corollary 2.13 tells us that all full models have this form, perhaps modulo a bilogical morphism; this may be seen as a justification of the use of the term *full* to describe the notion of full model.

Given an abstract logic $\mathbb{L}$, consider the projection of A onto $A^* = A/\widetilde{\mathit{\Omega}}(\mathbb{L})$. It is a particular case of the situation described in Proposition 1.20, which tells us that the $\mathcal{S}$-filters on A^* are the result of reducing the $\mathcal{S}$-filters F on A such that $\widetilde{\mathit{\Omega}}(\mathbb{L})$ is compatible with F, that is, such that $\widetilde{\mathit{\Omega}}(\mathbb{L}) \subseteq \mathit{\Omega}_A(F)$. Then we obtain the next characterization, which is particularly interesting for it offers another view of the "fullness" of full models: An abstract logic $\mathbb{L}$ is a full model of $\mathcal{S}$ if and only if its closure system consists of *all* the $\mathcal{S}$-filters that correspond to an $\mathcal{S}$-filter on the reduction A^* of A by $\widetilde{\mathit{\Omega}}(\mathbb{L})$.

THEOREM 2.14. *An abstract logic $\mathbb{L} = \langle A, \mathcal{C} \rangle$ is a full model of $\mathcal{S}$ if and only if $\mathcal{C} = \{ F \in \mathcal{F}i_\mathcal{S} A : \widetilde{\mathit{\Omega}}_A(\mathcal{C}) \subseteq \mathit{\Omega}_A(F) \}$.*

PROOF. ($\Rightarrow$): If $\mathbb{L} = \langle A, \mathcal{C} \rangle$ is a full model of $\mathcal{S}$ and $F \in \mathcal{C}$, then by Proposition 2.9 $F \in \mathcal{F}i_\mathcal{S} A$, and in general $\widetilde{\mathit{\Omega}}_A(\mathcal{C}) \subseteq \mathit{\Omega}_A(F)$, by Proposition 1.2. In order to prove the other inclusion assume that $F \in \mathcal{F}i_\mathcal{S} A$ is such that $\widetilde{\mathit{\Omega}}_A(\mathcal{C}) \subseteq \mathit{\Omega}_A(F)$, that is, $\widetilde{\mathit{\Omega}}_A(\mathcal{C})$ is compatible with F. By Proposition 1.20 there is some $G \in \mathcal{F}i_\mathcal{S}\big(A/\widetilde{\mathit{\Omega}}_A(\mathcal{C})\big)$ such that $F = \pi^{-1}[G]$, where π is the projection from A onto $A/\widetilde{\mathit{\Omega}}_A(\mathcal{C})$. Since π is a bilogical morphism from $\mathbb{L}$ onto $\mathbb{L}^*$, and $\mathcal{C}^* = \mathcal{F}i_\mathcal{S}\big(A/\widetilde{\mathit{\Omega}}_A(\mathcal{C})\big)$ because $\mathbb{L}$ is a full model of $\mathcal{S}$, it results that $F \in \mathcal{C}$.

($\Leftarrow$): Assume now that $\mathcal{C} = \{ F \in \mathcal{F}i_\mathcal{S} A : \widetilde{\mathit{\Omega}}_A(\mathcal{C}) \subseteq \mathit{\Omega}_A(F) \}$. Using Proposition 1.20 again, we see that π is a bilogical morphism between $\mathbb{L} = \langle A, \mathcal{C} \rangle$

and the abstract logic $\langle A/\widetilde{\Omega}_A(\mathcal{C}), \mathcal{F}i_{\mathcal{S}}(A/\widetilde{\Omega}_A(\mathcal{C}))\rangle$. From this it follows that $\mathcal{C}^* = \mathcal{F}i_{\mathcal{S}}(A/\widetilde{\Omega}_A(\mathcal{C}))$ and as a consequence $\mathbb{L}$ is a full model of $\mathcal{S}$.   ⊣

PROPOSITION 2.15. *An abstract logic $\mathbb{L}$ is a full model of $\mathcal{S}$ if and only if there is a full model of $\mathcal{S}$, $\mathbb{L}_\kappa$, on a formula algebra $\boldsymbol{Fm}_\kappa$ of suitable cardinality, and there is some $\theta \in \mathrm{Con}\,\boldsymbol{Fm}_\kappa$ such that $\mathbb{L}$ is isomorphic to $\mathbb{L}_\kappa/\theta$. And $\mathbb{L}$ is a reduced full model of $\mathcal{S}$ iff $\mathbb{L}$ is isomorphic to the reduction of a full model of $\mathcal{S}$ on $\boldsymbol{Fm}_\kappa$.*

PROOF. Simply repeat the construction in the proof of Proposition 1.16 and apply Proposition 2.11. The second part follows from the first and Proposition 1.13.   ⊣

2.2. $\mathcal{S}$-algebras

From the previous properties it follows that the reduced full models of $\mathcal{S}$ are exactly all those abstract logics of the form $\langle A, \mathcal{F}i_{\mathcal{S}}A\rangle$ which are reduced. This observation suggests that we should highlight the algebras for which this situation happens:

DEFINITION 2.16. *If $\mathcal{S}$ is a sentential logic, then an algebra A is an $\mathcal{S}$-algebra if and only if the abstract logic $\langle A, \mathcal{F}i_{\mathcal{S}}A\rangle$ is reduced, that is, iff it is the algebraic reduct of a reduced full model of $\mathcal{S}$.*
The class of all $\mathcal{S}$-algebras will be denoted by $\mathbf{Alg}\mathcal{S}$.

Thus the Lindenbaum-Tarski algebra $\boldsymbol{Fm}^* = \boldsymbol{Fm}/\widetilde{\Omega}(\mathcal{S})$ is an $\mathcal{S}$-algebra as well. The term "$\mathcal{S}$-algebra" has already been used in the literature, in some algebraic approaches to smaller classes of sentential logics, to denote a class of algebras naturally associated with a sentential logic $\mathcal{S}$. Perhaps the most well-known case is Rasiowa [1974], where this term, introduced in Rasiowa and Sikorski [1953], is used for a class of logics of implicative character, the so-called *standard systems of implicative extensional propositional calculi*. In Czelakowski [1980] Proposition 8.5 it is proved that in all these cases Rasiowa's "$\mathcal{S}$-algebras" are the algebraic reducts of their reduced matrices; as we shall see, our Proposition 3.2 will confirm that the class we call $\mathcal{S}$-algebras coincides with the class she called by this name. The first extension of this terminology was performed in Czelakowski [1981] to *equivalential logics with an algebraic semantics*, a larger class of logics that also falls under the scope of Proposition 3.2.

From the definition and previous results on full models we immediately have:

PROPOSITION 2.17. *For any abstract logic* $\mathbb{L} = \langle A, \mathcal{C} \rangle$ *the following conditions are equivalent:*

(i) $\mathbb{L}$ *is a reduced full model of* $\mathcal{S}$.
(ii) $\mathbb{L}$ *is reduced and* $\mathcal{C} = \mathcal{F}i_{\mathcal{S}} A$.
(iii) A *is an* $\mathcal{S}$-algebra and $\mathcal{C} = \mathcal{F}i_{\mathcal{S}} A$. $\dashv$

PROPOSITION 2.18. *Let* $\mathbb{L} = \langle A, \mathrm{C} \rangle$ *be a full model of* $\mathcal{S}$. *Then the algebra* A^* *is an* $\mathcal{S}$-algebra, and so $\widetilde{\Omega}(\mathbb{L}) \in \mathrm{Con}_{\mathsf{Alg}\mathcal{S}} A$. $\dashv$

It may be interesting to observe that in order to obtain the class of $\mathcal{S}$-algebras one does not need the notion of full model; the notion of model is enough:

PROPOSITION 2.19. *For any sentential logic* $\mathcal{S}$, *the class of* $\mathcal{S}$-algebras is the class of the algebraic reducts of all the reduced models of $\mathcal{S}$.

PROOF. For any A, the abstract logic $\langle A, \mathcal{F}i_{\mathcal{S}} A \rangle$ is a model of $\mathcal{S}$, and if $A \in$ **Alg**$\mathcal{S}$ then it is reduced. Conversely, if $\mathbb{L} = \langle A, \mathcal{C} \rangle$ is any model of $\mathcal{S}$ then $\mathcal{C} \subseteq \mathcal{F}i_{\mathcal{S}} A$ by 2.7, therefore $\widetilde{\Omega}(\langle A, \mathcal{F}i_{\mathcal{S}} A \rangle) \subseteq \widetilde{\Omega}(\mathbb{L})$; thus if $\mathbb{L}$ is reduced then so is $\langle A, \mathcal{F}i_{\mathcal{S}} A \rangle$, and this means that $A \in$ **Alg**$\mathcal{S}$. $\dashv$

PROPOSITION 2.20. *The class of* $\mathcal{S}$-algebras is closed under isomorphisms.

PROOF. If A_1 and A_2 are two isomorphic algebras, then it is easy to prove, using 1.19, that the lattices $\mathcal{F}i_{\mathcal{S}} A_1$ and $\mathcal{F}i_{\mathcal{S}} A_2$ are also isomorphic by the induced mapping; therefore by 1.21 the abstract logics $\langle A_1, \mathcal{F}i_{\mathcal{S}} A_1 \rangle$ and $\langle A_2, \mathcal{F}i_{\mathcal{S}} A_2 \rangle$ are isomorphic abstract logics. Hence one of them is reduced iff the other one is. Therefore A_1 is an $\mathcal{S}$-algebra iff A_2 is. $\dashv$

Although it contains some redundancies, the next result is of interest since it has a general form corresponding to many of Verdú's results for particular $\mathcal{S}$, especially those in Font and Verdú [1988], [1989b], [1991] and those in Verdú [1978], [1979], [1987]. See Chapter 5 for the exact correspondence between 2.21 and these particular results; as we show there, using 2.21, these particular results give nice characterizations of full models of $\mathcal{S}$ in many cases where the $\mathcal{S}$-algebras and the $\mathcal{S}$-filters on them have already been characterized.

PROPOSITION 2.21. *For any abstract logic* $\mathbb{L} = \langle A, \mathrm{C} \rangle$ *the following conditions are equivalent:*

(i) $\mathbb{L}$ *is a full model of* $\mathcal{S}$.
(ii) $A / \widetilde{\Omega}(\mathbb{L})$ *is an* $\mathcal{S}$-algebra and $\mathcal{C}/\widetilde{\Omega}(\mathbb{L}) = \mathcal{F}i_{\mathcal{S}}(A/\widetilde{\Omega}(\mathbb{L}))$.
(iii) *There is a bilogical morphism between the abstract logic* $\mathbb{L}$ *and an abstract logic* $\mathbb{L}' = \langle A', \mathrm{C}' \rangle$ *such that* A' *is an* $\mathcal{S}$-algebra and $\mathcal{C}' = \mathcal{F}i_{\mathcal{S}} A'$. $\dashv$

Notice also that the characterizations of algebras in terms of closure operators contained in Verdú [1985], having the form "an algebra belongs to such-and-such class of algebras if and only if there is a closure operator on it satisfying some list of properties and being reduced", will become instances of the definition of S-algebra for those S whose full models are characterized by that list of properties.

THEOREM 2.22 (Completeness Theorem). *Any sentential logic S is complete with respect to the following classes of abstract logics:*

(1) *The class of all full models of S.*
(2) *The class of all abstract logics of the form $\langle A, \mathcal{F}i_S A \rangle$ for all algebras A.*
(3) *The class of all reduced full models of S, i.e., the class of all abstract logics of the form $\langle A, \mathcal{F}i_S A \rangle$ for all $A \in \mathbf{Alg}S$.*

PROOF. The three classes consist of models of S, and S^* belongs to all of them. Therefore these classes satisfy the conditions of Proposition 2.6, so S is complete with respect to each one of them.   ⊣

The usefulness of this result, especially its part (3), for a particular S, depends on the characterizations we may have of the class $\mathbf{Alg}S$ and of the operator of S-filter-generation on the algebras of this class.

We have seen that the relationship between $\mathbf{Alg}S$ and $\mathbf{FMod}S$ is similar to the one existing between $\mathbf{Alg}^*S$ and $\mathbf{Matr}S$: in both cases the algebras are the algebraic reducts of the reduced models under consideration. Now we determine the precise relationship between the two classes of algebras.

THEOREM 2.23. *For any sentential logic S, the class $\mathbf{Alg}S$ is the class of all subdirect products of algebras in the class $\mathbf{Alg}^*S$; in symbols: $\mathbf{Alg}S = \mathrm{P}_{\mathrm{SD}}\mathbf{Alg}^*S$.*

PROOF. If $A \in \mathbf{Alg}S$, we have that $Id_A = \widetilde{\Omega}_A(\mathcal{F}i_S A) = \bigcap\{\Omega_A(F) : F \in \mathcal{F}i_S A\}$. In this situation we know that A is a subdirect product of the quotients $\{A/\Omega_A(F) : F \in \mathcal{F}i_S A\}$, and it is always true that $A/\Omega_A(F) \in \mathbf{Alg}^*S$ when $F \in \mathcal{F}i_S A$. Conversely, let A be a subdirect product of a family $\{A_i : i \in I\} \subseteq \mathbf{Alg}^*S$; thus for each $i \in I$ there is some $F_i \in \mathcal{F}i_S A_i$ such that $\Omega_{A_i}(F_i) = Id_{A_i}$. Now consider the closure system $\mathcal{C}$ generated on A by the family of subsets $\{\pi_i^{-1}[F_i] : i \in I\}$, where π_i is the canonical epimorphism from A onto A_i. The abstract logic $\langle A, \mathcal{C} \rangle$ is obviously a model of S, and it is reduced: If $\langle a, b \rangle \in \widetilde{\Omega}_A(\mathcal{C}) = \bigcap\{\Omega_A(T) : T \in \mathcal{C}\}$ then for each $i \in I$, $\langle a, b \rangle \in \Omega_A(\pi_i^{-1}[F_i]) = \pi_i^{-1}[\Omega_{A_i}(F_i)] = \pi_i^{-1}[Id_{A_i}] = \ker \pi_i$, that is, $\pi_i(a) = \pi_i(b)$ for all $i \in I$, which implies $a = b$. We have proved that $\langle A, \mathcal{C} \rangle$ is a reduced model of S. By Proposition 2.19, $A \in \mathbf{Alg}S$.   ⊣

COROLLARY 2.24. *For any sentential logic S, $\mathbf{Alg}^*S \subseteq \mathbf{Alg}S$; and $\mathbf{Alg}^*S =$* $\mathbf{Alg}S$ *if and only if the class $\mathbf{Alg}^*S$ is closed under subdirect products.*   ⊣

Among the many consequences of Theorem 2.23 is that the class $\mathbf{Alg}S$ is always closed under subdirect products; hence it is also closed under direct products. Since quasivarieties are always closed under subdirect products, it may be interesting to record the following:

COROLLARY 2.25. *If the class $\mathbf{Alg}^*S$ is a quasivariety, then $\mathbf{Alg}^*S = \mathbf{Alg}S$.* *In particular, this holds when $\mathbf{Alg}^*S$ is a variety.*   ⊣

This covers many of the common sentential logics, whose associated classes of algebras are quasivarieties or varieties. Moreover, in Chapter 3 we prove that for all protoalgebraic sentential logics, a rather wide class, the equality $\mathbf{Alg}^*S =$ $\mathbf{Alg}S$ also holds, even if this class is not a variety or a quasi-variety; the logic LJ of the "last judgement" invented by Herrmann [1993b] is an example where this class is not even elementary. In addition, the converse of Corollary 2.25 is not true, that is, $\mathbf{Alg}S$ can be a quasivariety, or even a variety, without being equal to $\mathbf{Alg}^*S$; again the $\{\wedge, \vee\}$-fragment of classical logic is an example, see Section 5.1.1.

Another consequence of Theorem 2.23 is that, even if they are different, these two classes generate the same quasivariety, and a fortiori the same variety:

PROPOSITION 2.26. *For each sentential logic S, the classes of algebras $\mathbf{Alg}S$* *and $\mathbf{Alg}^*S$ generate the same quasivariety and the same variety; this variety is* *the class $\mathbf{K}_S$, that is, the variety generated by the Lindenbaum-Tarski algebra* Fm^*.

PROOF. From the result in Theorem 2.23 it follows that the quasivariety generated by $\mathbf{Alg}S$ is included in the quasivariety generated by $\mathbf{Alg}^*S$; but since by Corollary 2.24 $\mathbf{Alg}^*S \subseteq \mathbf{Alg}S$, the opposite inclusion also holds, and the two quasivarieties are equal. As a consequence, the varieties generated by them also coincide, and by Proposition 1.23 this variety is $\mathbf{K}_S$.   ⊣

This result adds further support, from within our theory, to the common idea that if one insists on associating a variety with every sentential logic in a uniform way, then the variety $\mathbf{K}_S$ generated by the Lindenbaum-Tarski algebra is the most natural one; but we have already mentioned that there are cases where there is no point in doing so.

PROPOSITION 2.27. *If S and S' are two sentential logics, and S' is stronger* *than S, then $\mathbf{Alg}S' \subseteq \mathbf{Alg}S$ and $\mathbf{Alg}^*S' \subseteq \mathbf{Alg}^*S$.*

PROOF. If $\mathcal{S} \leqslant \mathcal{S}'$ then for any A, $\mathcal{F}i_{\mathcal{S}'} A \subseteq \mathcal{F}i_{\mathcal{S}} A$, and this directly implies $\mathbf{Alg^*}\mathcal{S}' \subseteq \mathbf{Alg^*}\mathcal{S}$. Then, using Theorem 2.23, this implies $\mathbf{Alg}\mathcal{S}' \subseteq \mathbf{Alg}\mathcal{S}$. ⊣

So far we have associated *three classes of algebras* with an arbitrary sentential logic: $\mathbf{Alg^*}\mathcal{S}$, $\mathbf{Alg}\mathcal{S}$, $\mathbf{K}_{\mathcal{S}}$. We have seen that $\mathbf{Alg^*}\mathcal{S} \subseteq \mathbf{Alg}\mathcal{S} \subseteq \mathbf{K}_{\mathcal{S}}$; the two extreme ones have already been considered in the literature, but sometimes $\mathbf{Alg^*}\mathcal{S}$ is too small and $\mathbf{Alg^*}\mathcal{S} \subsetneq \mathbf{Alg}\mathcal{S} = \mathbf{K}_{\mathcal{S}}$, and sometimes $\mathbf{K}_{\mathcal{S}}$ is too large and $\mathbf{Alg^*}\mathcal{S} = \mathbf{Alg}\mathcal{S} \subsetneq \mathbf{K}_{\mathcal{S}}$, as in several examples we have already mentioned. It is then natural to ask the following question:

OPEN PROBLEM. *Is there a sentential logic $\mathcal{S}$ such that the three classes of algebras are all different, that is, such that $\mathbf{Alg^*}\mathcal{S} \subsetneq \mathbf{Alg}\mathcal{S} \subsetneq \mathbf{K}_{\mathcal{S}}$?*[17]

By Proposition 3.2, a logic with this property cannot be protoalgebraic, and by Corollary 2.25 the first two classes cannot be quasivarieties. Moreover, by the results we will find in Chapter 4, such a logic cannot be selfextensional and at the same time satisfy the Property of Conjunction, or the Deduction Theorem.

2.3. The lattice of full models over an algebra

In this section we will prove that the ordered set $\langle \mathcal{F}\mathcal{M}od_{\mathcal{S}} A, \leqslant \rangle$ and the ordered set $\langle \mathrm{Con}_{\mathbf{Alg}\mathcal{S}} A, \subseteq \rangle$ are isomorphic through the Tarski operator (Theorem 2.30) and that the second one is a complete lattice (Theorem 2.31); as a consequence the set of all full models of $\mathcal{S}$ over an algebra will also become a complete lattice. We begin by introducing a construction which will turn out to be inverse to the Tarski operator, and which has an interest of its own.

DEFINITION 2.28. *Let A be any algebra. For any $\theta \in \mathrm{Con}\, A$, we denote by $\widetilde{\mathbf{H}}_A(\theta) = \langle A, C_\theta \rangle$ the abstract logic projectively generated on A from the abstract logic $\langle A/\theta, \mathcal{F}i_{\mathcal{S}}(A/\theta) \rangle$ by the canonical projection π of A onto A/θ.*

Note that with this definition π becomes a bilogical morphism between $\widetilde{\mathbf{H}}_A(\theta)$ and the abstract logic $\langle A/\theta, \mathcal{F}i_{\mathcal{S}}(A/\theta) \rangle$. Now we record some general properties of this construction.

LEMMA 2.29. *For any $\theta \in \mathrm{Con}\, A$, it holds that $\theta \in \mathrm{Con}\, \widetilde{\mathbf{H}}_A(\theta)$. Moreover, it holds that $\widetilde{\mathbf{H}}_A(\theta)/\theta = \langle A/\theta, \mathcal{F}i_{\mathcal{S}}(A/\theta) \rangle$, that $\widetilde{\mathbf{H}}_A(\theta) \in \mathcal{F}\mathcal{M}od_{\mathcal{S}} A$ and that the mapping $\theta \mapsto \widetilde{\mathbf{H}}_A(\theta)$ is order-preserving: If $\theta_1, \theta_2 \in \mathrm{Con}\, A$ are such that $\theta_1 \subseteq \theta_2$ then $\widetilde{\mathbf{H}}_A(\theta_1) \leqslant \widetilde{\mathbf{H}}_A(\theta_2)$.*

---

[17]This problem was solved in the affirmative in Bou [2001] in the context of the study of certain subintuitionistic logics, and, independently, in Babyonyshev [2003] by an ad-hoc construcion.

PROOF. If $\langle a, b \rangle \in \theta$ then $\pi(a) = \pi(b)$ and thus $\mathrm{Fi}_{\mathcal{S}}^{A/\theta}\big(\pi(a)\big) = \mathrm{Fi}_{\mathcal{S}}^{A/\theta}\big(\pi(b)\big)$, therefore $\pi^{-1}\big[\mathrm{Fi}_{\mathcal{S}}^{A/\theta}\big(\pi(a)\big)\big] = \pi^{-1}\big[\mathrm{Fi}_{\mathcal{S}}^{A/\theta}\big(\pi(b)\big)\big]$. But by construction we know that $\mathrm{C}_{\theta} = \pi^{-1} \circ \mathrm{Fi}_{\mathcal{S}}^{A/\theta} \circ \pi$; therefore we get $\mathrm{C}_{\theta}(a) = \mathrm{C}_{\theta}(b)$. Thus we have proved that $\theta \in \mathrm{Con}\,\widetilde{\mathbf{H}}_A(\theta)$. The second part of the statement comes directly from the construction. Moreover, since $\langle A/\theta, \mathcal{F}i_{\mathcal{S}}(A/\theta)\rangle$ is always a full model of $\mathcal{S}$, by 2.11 $\widetilde{\mathbf{H}}_A(\theta)$ is also a full model of $\mathcal{S}$. To prove the last part of the Lemma, take $\theta_1, \theta_2 \in \mathrm{Con}\,A$ and consider the natural projections $\pi_1 : A \to A/\theta_1$ and $\pi_2 : A \to A/\theta_2$. If moreover $\theta_1 \subseteq \theta_2$ then the mapping $j(a/\theta_1) = a/\theta_2$ is an epimorphism from A/θ_1 onto A/θ_2, and the following diagram

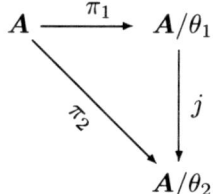

commutes. As a consequence, if $Z \in \mathcal{F}i_{\mathcal{S}}(A/\theta_2)$ then $j^{-1}[Z] \in \mathcal{F}i_{\mathcal{S}}(A/\theta_1)$. Due to this, the closure system of $\widetilde{\mathbf{H}}_A(\theta_2)$, which is projectively generated from $\mathcal{F}i_{\mathcal{S}}(A/\theta_2)$ by π_2, satisfies

$$
\begin{aligned}
\mathcal{C}_{\theta_2} &= \{\pi_2^{-1}[Z] : Z \in \mathcal{F}i_{\mathcal{S}}(A/\theta_2)\} = \\
&= \{\pi_1^{-1}\big[j^{-1}[Z]\big] : Z \in \mathcal{F}i_{\mathcal{S}}(A/\theta_2)\} \subseteq \\
&\subseteq \{\pi_1^{-1}[X] : X \in \mathcal{F}i_{\mathcal{S}}(A/\theta_1)\} = \mathcal{C}_{\theta_1},
\end{aligned}
$$

that is, it is contained in the closure system of $\widetilde{\mathbf{H}}_A(\theta_1)$, which is projectively generated from $\mathcal{F}i_{\mathcal{S}}(A/\theta_1)$ by π_1. Therefore, $\widetilde{\mathbf{H}}_A(\theta_1) \leqslant \widetilde{\mathbf{H}}_A(\theta_2)$ as was to be shown. $\dashv$

We will now prove[18] that, when restricted to $\mathrm{Con}_{\mathbf{Alg}\mathcal{S}}\,A$, this mapping is exactly the inverse of the Tarski operator, and that both mappings are order-isomorphisms.

THEOREM 2.30 (The Isomorphism Theorem). *For any algebra A, the Tarski operator $\widetilde{\Omega}_A$ is an order-isomorphism between the ordered sets $\langle \mathcal{FMod}_{\mathcal{S}}\,A, \leqslant\rangle$ and $\langle \mathrm{Con}_{\mathbf{Alg}\mathcal{S}}\,A, \subseteq\rangle$; and the mapping $\widetilde{\mathbf{H}}_A$ is its inverse.*

PROOF. As we have already observed in Proposition 2.18, if $\mathbb{L} \in \mathcal{FMod}_{\mathcal{S}}\,A$ then $\widetilde{\Omega}_A(\mathbb{L}) \in \mathrm{Con}_{\mathbf{Alg}\mathcal{S}}\,A$. Dually, in Lemma 2.29 we have seen that if $\theta \in$

---

[18]For an essentially different proof, see Font, Jansana, and Pigozzi [2006].

$\mathrm{Con}_{\mathbf{Alg}_{\mathcal{S}}} A$ then $\widetilde{\mathbf{H}}_{A}(\theta) \in \mathcal{F}Mod_{\mathcal{S}} A$. So both mappings are well-defined. Now we will see that they are bijections.

We first prove that $\widetilde{\mathbf{H}}_{A}\left(\widetilde{\mathbf{\Omega}}_{A}(\mathbb{L})\right) = \mathbb{L}$ assuming that $\mathbb{L} = \langle A, \mathcal{C} \rangle \in \mathcal{F}Mod_{\mathcal{S}} A$: If $\mathbb{L}$ is a full model of $\mathcal{S}$, we have already seen in 2.18 that A^{*} is an $\mathcal{S}$-algebra, and that $\widetilde{\mathbf{\Omega}}_{A}(\mathbb{L}) \in \mathrm{Con}_{\mathbf{Alg}_{\mathcal{S}}} A$; moreover, $\mathcal{C}$ is projectively generated from its reduction $\mathcal{C}^{*} = \mathcal{F}i_{\mathcal{S}} A^{*}$ by the canonical projection from A onto $A^{*} = A / \widetilde{\mathbf{\Omega}}_{A}(\mathbb{L})$. By Definition 2.28, this is exactly $\widetilde{\mathbf{H}}_{A}\left(\widetilde{\mathbf{\Omega}}_{A}(\mathbb{L})\right)$, therefore $\widetilde{\mathbf{H}}_{A}\left(\widetilde{\mathbf{\Omega}}_{A}(\mathbb{L})\right) = \mathbb{L}$.

Let now $\theta \in \mathrm{Con}_{\mathbf{Alg}_{\mathcal{S}}} A$; we now prove that $\widetilde{\mathbf{\Omega}}_{A}\left(\widetilde{\mathbf{H}}_{A}(\theta)\right) = \theta$: Observe that $\widetilde{\mathbf{\Omega}}_{A/\theta}\left(\mathcal{F}i_{\mathcal{S}}(A/\theta)\right) = Id_{A/\theta}$ precisely because $\theta \in \mathrm{Con}_{\mathbf{Alg}_{\mathcal{S}}} A$. Then, using Proposition 1.7, we have

$$\widetilde{\mathbf{\Omega}}_{A}\left(\widetilde{\mathbf{H}}_{A}(\theta)\right) = \widetilde{\mathbf{\Omega}}_{A}\left(\pi^{-1}\left[\mathcal{F}i_{\mathcal{S}}(A/\theta)\right]\right)$$
$$= \pi^{-1}\left[\widetilde{\mathbf{\Omega}}_{A/\theta}\left(\mathcal{F}i_{\mathcal{S}}(A/\theta)\right)\right]$$
$$= \pi^{-1}\left[Id_{A/\theta}\right] = \theta$$

Putting together the results of the last two paragraphs, we obtain that both mappings are bijections. Since by its own definition $\widetilde{\mathbf{\Omega}}_{A}$ is order-preserving, and by Lemma 2.29 $\widetilde{\mathbf{H}}_{A}$ is also order-preserving, we conclude that they are order-isomorphisms between the two ordered sets. This ends the proof of the theorem.   ⊣

Independently of this result, we can determine that the set $\mathrm{Con}_{\mathbf{Alg}_{\mathcal{S}}} A$ involved in the Isomorphism Theorem, ordered under $\subseteq$, has a lattice structure.

THEOREM 2.31. *For any algebra* A, *the ordered set* $\langle \mathrm{Con}_{\mathbf{Alg}_{\mathcal{S}}} A, \subseteq \rangle$ *is a complete lattice, where* inf *is intersection.*

PROOF. Let $\{\theta_{i} : i \in I\}$ be a non-empty family of elements of $\mathrm{Con}_{\mathbf{Alg}_{\mathcal{S}}} A$, and put $\theta = \bigcap\{\theta_{i} : i \in I\}$; we will prove that $\theta \in \mathrm{Con}_{\mathbf{Alg}_{\mathcal{S}}} A$. First of all we observe that for any $a \in A$, $a/\theta = \bigcap\{a/\theta_{i} : i \in I\}$, and consider, for each $i \in I$, the mapping $h_{i} : A/\theta \to A/\theta_{i}$ defined by $h_{i}(a/\theta) = a/\theta_{i}$, which is an epimorphism. By assumption, for every $i \in I$, the abstract logic $\mathbb{L}_{i} = \langle A/\theta_{i}, \mathcal{F}i_{\mathcal{S}}(A/\theta_{i}) \rangle$ is reduced, and we have to show that $\mathbb{L} = \langle A/\theta, \mathcal{F}i_{\mathcal{S}}(A/\theta) \rangle$ is reduced. Since $h_{i}^{-1}\left[\mathcal{F}i_{\mathcal{S}}(A/\theta_{i})\right] \subseteq \mathcal{F}i_{\mathcal{S}}(A/\theta)$ by Proposition 1.7 we have

$$\widetilde{\mathbf{\Omega}}(\mathbb{L}) \subseteq \widetilde{\mathbf{\Omega}}_{A/\theta}\left(h_{i}^{-1}\left[\mathcal{F}i_{\mathcal{S}}(A/\theta_{i})\right]\right) =$$
$$= h_{i}^{-1}\left[\widetilde{\mathbf{\Omega}}_{A/\theta_{i}}\left(\mathcal{F}i_{\mathcal{S}}(A/\theta_{i})\right)\right] = h_{i}^{-1}\left[Id_{A/\theta_{i}}\right]$$

because $\mathbb{L}_{i}$ is reduced. Therefore if $\langle a/\theta, b/\theta \rangle \in \widetilde{\mathbf{\Omega}}(\mathbb{L})$ then $a/\theta_{i} = b/\theta_{i}$ for each $i \in I$, and as a consequence $a/\theta = b/\theta$. This proves that $\mathbb{L}$ is reduced, that is, that $\theta \in \mathrm{Con}_{\mathbf{Alg}_{\mathcal{S}}} A$. Thus $\mathrm{Con}_{\mathbf{Alg}_{\mathcal{S}}} A$ is closed under intersections of non-empty families. On the other hand, if A is trivial (1-element) then either

$\mathcal{F}i_\mathcal{S}A = \{A\}$, if $\mathcal{S}$ has theorems, or $\mathcal{F}i_\mathcal{S}A = \{\emptyset, A\}$, if $\mathcal{S}$ doesn't; in either case the abstract logic $\langle A, \mathcal{F}i_\mathcal{S}A \rangle$ is reduced, and hence it is a full model of $\mathcal{S}$, which shows that $A \in \mathsf{Alg}\mathcal{S}$. As a consequence, for an arbitrary A, the universal congruence $A \times A \in \mathrm{Con}_{\mathsf{Alg}\mathcal{S}}A$, which concludes the proof that the ordered set $\mathrm{Con}_{\mathsf{Alg}\mathcal{S}}A$ is a complete lattice. ⊣

Since Theorem 2.30 establishes an order-isomorphism, we get immediately:

COROLLARY 2.32. *For any A, the ordered set $\langle \mathcal{F}\!Mod_\mathcal{S}A, \leqslant \rangle$ is a complete lattice, and the Tarski operator is a lattice isomorphism between $\langle \mathcal{F}\!Mod_\mathcal{S}A, \leqslant \rangle$ and $\langle \mathrm{Con}_{\mathsf{Alg}\mathcal{S}}A, \subseteq \rangle$.* ⊣

Note that, although $\mathcal{F}\!Mod_\mathcal{S}A$ is a subset of the complete lattice of all abstract logics over A, it need not be a sublattice; indeed, we do not have nice characterizations of the lattice operations in $\langle \mathcal{F}\!Mod_\mathcal{S}A, \leqslant \rangle$. The only thing we can say is that, as a consequence of the preceding results, given any collection $\{\mathbb{L}_i : i \in I\}$ of full models of $\mathcal{S}$ on the same algebra A, its infimum in the lattice of full models of $\mathcal{S}$ can be obtained as the abstract logic projectively generated from $\langle A/\theta, \mathcal{F}i_\mathcal{S}(A/\theta) \rangle$ by the canonical projection of A onto A/θ, where $\theta = \bigcap \{\widetilde{\boldsymbol{\Omega}}(\mathbb{L}_i) : i \in I\}$.

PROPOSITION 2.33. *Let $\mathbb{L}_1$ and $\mathbb{L}_2$ be two full models of $\mathcal{S}$, and let h be a bilogical morphism between them. Then the mapping $\mathcal{C} \mapsto \{h[X] : X \in \mathcal{C}\}$ is an isomorphism between the lattice of all full models of $\mathcal{S}$ on A_1 extending $\mathbb{L}_1$ and the lattice of all full models of $\mathcal{S}$ on A_2 extending $\mathbb{L}_2$. And also the principal ideals of $\mathrm{Con}_{\mathsf{Alg}\mathcal{S}}A_1$ and of $\mathrm{Con}_{\mathsf{Alg}\mathcal{S}}A_2$ determined respectively by $\widetilde{\boldsymbol{\Omega}}_{A_1}(\mathbb{L}_1)$ and by $\widetilde{\boldsymbol{\Omega}}_{A_2}(\mathbb{L}_2)$ are isomorphic.*

PROOF. In Corollary 1.6 we have seen that the mapping $\mathcal{C} \mapsto \hat{h}(\mathcal{C}) = \{h[X] : X \in \mathcal{C}\}$ is an isomorphism between the lattices of all abstract logics on A_1 extending $\mathbb{L}_1$ and of all abstract logics on A_2 extending $\mathbb{L}_2$. But this mapping establishes in each case a bilogical morphism between the two abstract logics whose closure systems are $\mathcal{C}$ and $\hat{h}(\mathcal{C})$, and by Proposition 2.11 one of these is a full model of $\mathcal{S}$ if and only if the other one is. ⊣

And as a particular case we have:

COROLLARY 2.34. *If A, B are algebras and $h : A \to B$ is an epimorphism satisfying any of the equivalent conditions appearing in Proposition 1.21, then h induces an isomorphism between the complete lattices $\mathcal{F}\!Mod_\mathcal{S}A$ and $\mathcal{F}\!Mod_\mathcal{S}B$; and also the lattices $\mathrm{Con}_{\mathsf{Alg}\mathcal{S}}A$ and $\mathrm{Con}_{\mathsf{Alg}\mathcal{S}}B$ are isomorphic.*

PROOF. This is the conjunction of 1.21 and 2.33 taking into account that, by 2.10, the abstract logic $\langle A, \mathcal{F}i_\mathcal{S}A \rangle$ is the weakest full model of $\mathcal{S}$ on A. ⊣

Finally we will use the language of categories to express the fact that, in some sense, using S-algebras is "equivalent" to using reduced full models of S, and that the process of reduction $\mathbb{L} \longmapsto \mathbb{L}^*$ has a good behaviour when considered globally, as a relationship between two categories of abstract logics.

THEOREM 2.35. *The algebraic category of the S-algebras together with homomorphisms is isomorphic to the category whose objects are the reduced full models of S, and whose arrows are the logical morphisms between its objects.*

PROOF. It is trivial to check that the class of abstract logics mentioned in the statement is really a category, since the identity mapping is a logical morphism, the composition of two logical morphisms is a logical morphism, and this composition is associative. To see that the category of S-algebras is isomorphic to the category of reduced full models of S it is enough to consider the functor defined on objects by $A \longmapsto \langle A, \mathcal{F}i_S A \rangle$, and defined on arrows by the identity: We know that if $A \in \mathbf{Alg}S$ then $\langle A, \mathcal{F}i_S A \rangle \in \mathbf{FMod}^*S$, and that every reduced full model of S is of this form, so this is a bijection between objects; and since for every $h \in \mathrm{Hom}(A, B)$ and every $F \in \mathcal{F}i_S B$, $h^{-1}[F] \in \mathcal{F}i_S A$, h is a logical morphism between $\langle A, \mathcal{F}i_S A \rangle$ and $\langle B, \mathcal{F}i_S B \rangle$, thus clearly this is a functor at the arrows level, and this finishes the proof that this functor is an isomorphism between the two categories. $\dashv$

The category of reduced full models of S considered in 2.35 is trivially a full subcategory of the category whose objects are all full models of S with logical morphisms as arrows. But if we only use surjective arrows then we obtain a more precise relationship between both categories of abstract logics:

THEOREM 2.36. *The category $\mathfrak{L}^*$ of reduced full models of S with surjective logical morphisms is a full reflective subcategory of the category $\mathfrak{L}$ of all full models of S with surjective logical morphisms; and the reflector is the functor associated with the process of "reduction": $\mathbb{L} \longmapsto \mathbb{L}^*$.*

PROOF. $\mathfrak{L}^*$ is trivially a full subcategory of $\mathfrak{L}$. In order to check that the process of reduction $\mathbb{L} \mapsto \langle \mathbb{L}^*, \pi_{\mathbb{L}} \rangle$ (where $\pi_{\mathbb{L}} : \mathbb{L} \to \mathbb{L}^*$ is the canonical projection) gives the announced reflector, it is enough to check (see Balbes and Dwinger [1974] I.18.2, for instance) that for an arbitrary surjective logical morphism $f : \mathbb{L} \to \mathbb{L}'$ between an $\mathbb{L} \in \mathbf{FMod}S$ and an $\mathbb{L}' \in \mathbf{FMod}^*S$ there is a unique surjective logical morphism $f^* : \mathbb{L}^* \to \mathbb{L}'$ such that $f^* \circ \pi_{\mathbb{L}} = f$. Since $\pi_{\mathbb{L}}$ is a bilogical morphism, we can use Proposition 1.15 if we prove that $\ker f \supseteq \ker \pi_{\mathbb{L}} = \widetilde{\Omega}(\mathbb{L})$. For this, consider the logic $\mathbb{L}_0$ projectively generated from $\mathbb{L}'$ by f; since f is an epimorphism, it becomes a bilogical morphism between $\mathbb{L}_0$ and $\mathbb{L}'$, and since by assumption $f^{-1}[T] \in \mathcal{C}$ for all $T \in \mathcal{C}'$, it results that $\mathbb{L} \leqslant \mathbb{L}_0$. Now, using Proposition 1.7 and that $\mathbb{L}'$ is reduced, we have that $\ker \pi_{\mathbb{L}} = \widetilde{\Omega}(\mathbb{L}) \subseteq \widetilde{\Omega}(\mathbb{L}_0) = f^{-1}[\widetilde{\Omega}(\mathbb{L}')] = f^{-1}[Id_{A'}] = \ker f$. Then we can use Proposition 1.15 to obtain a unique logical morphism $f^* : \mathbb{L}^* \to \mathbb{L}'$ with $f^* \circ \pi_{\mathbb{L}} = f$; and this equality implies it is surjective. $\dashv$

In Theorem 2.44 we will find a better result for a restricted class of sentential logics, where this reflector will reflect all logical morphisms, and not just the surjective ones.

2.4. Full models and metalogical properties

In this section we will see how some typical metalogical properties are inherited by full models of a sentential logic, while others may require additional assumptions. We have already noted in Proposition 2.9 that every full model of $\mathcal{S}$ inherits some of the basic properties of a sentential logic $\mathcal{S}$: those of being finitary, of having theorems and of not having theorems. Clearly the second of these properties is inherited by arbitrary models, while it is easy to see that the first and the third one are not.

In general, the metalogical properties under consideration must be such that it makes sense to ask whether an arbitrary abstract logic satisfies them. That is, they must be properties of the closure operator $\mathrm{Cn}_\mathcal{S}$ associated with $\vdash_\mathcal{S}$ and of its relationship with the algebraic structure of the underlying algebra. Most of them can be expressed in the form of a Gentzen-style rule for the derivability relation $\vdash_\mathcal{S}$. In order to obtain a useful degree of precision we give the following definition of Gentzen-style rule. A **sequent** will be a pair $\langle \Gamma, \varphi \rangle$, written $\Gamma \vdash \varphi$, where Γ is a finite set of formulas and φ is a formula. A **Gentzen-style rule** is a pair which consists of a finite set $\{\Gamma_i \vdash \varphi_i : i < k\}$ of sequents and a sequent $\Gamma \vdash \varphi$, which follows from the set according to the rule; the rule is often writen symbolycally in the "fraction" form

$$\frac{\{\Gamma_i \vdash \varphi_i : i < k\}}{\Gamma \vdash \varphi} \, , \tag{2.7}$$

and one says that a sentential logic $\mathcal{S}$ **satisfies the Gentzen-style rule** represented in (2.7) whenever for any substitution σ the following implication holds:

If for all $i < k$, $\sigma[\Gamma_i] \vdash_\mathcal{S} \sigma(\varphi_i)$ holds, then $\sigma[\Gamma] \vdash_\mathcal{S} \sigma(\varphi)$ holds.   (2.8)

In practice Gentzen-style rules are often described by using *schemes* that group together rules having a common form. For example, the expression

$$\frac{\Gamma, \psi_1 \vdash \varphi \qquad \Gamma, \psi_2 \vdash \varphi}{\Gamma, \psi_1 \vee \psi_2 \vdash \varphi} \tag{2.9}$$

has to be understood as varying over all finite sets of formulas Γ and all formulas φ, ψ_1, ψ_2. Strictly speaking it describes an infinite set of Gentzen-style rules which is closed under substitution instances (i.e. if it contains a rule, it contains the rule we obtain by applying an arbitrary substitution to all its formulas). In this way one does not need to use substitutions when characterizing the sentential logics that satisfy it, and we have that the sentence "(2.9) is a (Gentzen-style) rule of $\mathcal{S}$" actually means that for all finite Γ and all φ, ψ_1, ψ_2, if $\Gamma, \psi_1 \vdash_{\mathcal{S}} \varphi$ and $\Gamma, \psi_2 \vdash_{\mathcal{S}} \varphi$ then $\Gamma, \psi_1 \vee \psi_2 \vdash_{\mathcal{S}} \varphi$, as the rule scheme suggests.

Several of the properties considered in this section are of this kind, and one of the ways of further formalizing these issues in a general setting is by the use of Gentzen systems; we do this in Chapter 4.

By contrast, by a ***Hilbert-style rule*** of $\mathcal{S}$ we mean any sequent $\Gamma \vdash \varphi$ such that $\Gamma \vdash_{\mathcal{S}} \varphi$ holds. It is clear that both Hilbert-style and Gentzen-style rules can be formulated for an abstract logic $\mathbb{L} = \langle A, \mathrm{C} \rangle$ by substituting the $\vdash_{\mathcal{S}}$ relation by the closure operator C of $\mathbb{L}$, in an obvious way. Hence, they are metalogical properties of a sentential logic suitable to be investigated in the sense explained above. Note that, actually, an abstract logic is a model of $\mathcal{S}$ iff it satisfies all Hilbert-style rules of $\mathcal{S}$.

The congruence property

Recall that, for an arbitrary closure operator C, we denote by C^T the closure operator whose closure system is $\mathcal{C}^T = \{T' \in \mathcal{C} : T \subseteq T'\}$. We introduce an equivalence relation and a mapping naturally associated with any closure operator:

DEFINITION 2.37. *Let* C *be a closure operator on a set* A. *Then the **Frege relation** of* C *is:*

$$\Lambda(\mathrm{C}) = \big\{ \langle a, b \rangle \in A \times A : \mathrm{C}(a) = \mathrm{C}(b) \big\}.$$

*The **Frege operator** is the mapping* $\Lambda_{\mathrm{C}} : F \subseteq A \longmapsto \Lambda_{\mathrm{C}}(F) = \Lambda(\mathrm{C}^F)$.

If $\mathbb{L} = \langle A, \mathrm{C} \rangle$ *is an abstract logic, it will be convenient to use the notations* $\Lambda(\mathbb{L})$ *and* $\Lambda_{\mathbb{L}}$ *instead of* $\Lambda(\mathrm{C})$ *and* Λ_{C} *respectively.*

Note that $\Lambda(\mathrm{C}) = \Lambda_{\mathrm{C}}(\mathrm{C}(\emptyset))$, and that Λ_{C} is always order-preserving: if $F \subseteq G$ then $\Lambda_{\mathrm{C}}(F) \subseteq \Lambda_{\mathrm{C}}(G)$. Moreover:

PROPOSITION 2.38. *A closure operator* C *on a set* A *is finitary iff the Frege operator* Λ_C *preserves unions of directed families of subsets of* A; *that is, for any directed family* $\mathcal{D}$ *of subsets of* A, $\Lambda_C(\bigcup\mathcal{D}) = \bigcup\{\Lambda_C(F) : F \in \mathcal{D}\}$.

PROOF. Let $\mathcal{D}$ be any directed family of subsets of A. Since Λ_C is always order-preserving, we have $\Lambda_C(\bigcup\mathcal{D}) \supseteq \bigcup\{\Lambda_C(F) : F \in \mathcal{D}\}$. To prove the converse inclusion, suppose that $\langle a, b\rangle \in \Lambda_C(\bigcup\mathcal{D})$, that is, $C(\bigcup\mathcal{D}, a) = C(\bigcup\mathcal{D}, b)$. The finitarity of C implies that there are $c_1, \dots, c_n \in \bigcup\mathcal{D}$ such that

$$C(c_1, \dots, c_n, a) = C(c_1, \dots, c_n, b),$$

but since $\mathcal{D}$ is directed there is some $F \in \mathcal{D}$ such that all $c_i \in F$, which implies $C(F, a) = C(F, b)$, that is, $\langle a, b\rangle \in \Lambda_C(F)$, and therefore $\langle a, b\rangle \in \bigcup\{\Lambda_C(F) : F \in \mathcal{D}\}$. This proves that $\Lambda_C(\bigcup\mathcal{D}) = \bigcup\{\Lambda_C(F) : F \in \mathcal{D}\}$, and thus that Λ_C preserves unions of directed families of subsets of A. Conversely, take any nonempty $X \subseteq A$ and put $\mathcal{D} = \{F \subseteq X : F \text{ is finite }\}$; this is a directed family with $\bigcup\mathcal{D} = X$. If $a \in C(X)$ then for any $b \in X$ it holds that $C(X, a) = C(X, b) = C(X)$ so in particular $\langle a, b\rangle \in \Lambda_C(X) = \Lambda_C(\bigcup\mathcal{D}) = \bigcup\{\Lambda_C(F) : F \in \mathcal{D}\}$ by assumption. So there is a finite $F \subseteq X$ with $\langle a, b\rangle \in \Lambda_C(F)$, that is, $C(F, a) = C(F, b)$ which implies $a \in C(F, b)$. Since $F \cup \{b\}$ is a finite subset of X, this proves that C is finitary.   ⊣

For any closure operator C, the Frege relation $\Lambda(C)$ is trivially an equivalence relation. If $\mathbb{L} = \langle A, C\rangle$ is an abstract logic, then in general $\Lambda(\mathbb{L})$ is not a congruence of A; actually $\mathrm{Con}\mathbb{L} = \{\theta \in \mathrm{Con}A : \theta \subseteq \Lambda(\mathbb{L})\}$, and $\widetilde{\Omega}(\mathbb{L}) = \max \mathrm{Con}\mathbb{L}$ is precisely the greatest logical congruence of $\mathbb{L}$ included in $\Lambda(\mathbb{L})$.

DEFINITION 2.39. *We say that an abstract logic* $\mathbb{L} = \langle A, C\rangle$ *has the* **congruence property** *when* $\Lambda(\mathbb{L}) \in \mathrm{Con}A$, *that is, when* $\Lambda(\mathbb{L}) = \widetilde{\Omega}(\mathbb{L})$.

Note that a reduced abstract logic $\mathbb{L} = \langle A, C\rangle$ has the congruence property if and only if for all $a, b \in A$, $C(a) = C(b)$ implies $a = b$; that is, when $C(a) = C(b)$ holds exactly when $a = b$. By this, we see that properties of the underlying algebra can be expressed as properties of the closure operator, which can thus be extended to greater classes of abstract logics (for instance, if they are preserved under bilogical morphisms).

PROPOSITION 2.40. *The congruence property is preserved by bilogical morphisms. That is, if there is a bilogical morphism between two abstract logics then one of them has the congruence property if and only if the other one has it.*

PROOF. Suppose that h is a bilogical morphism between the abstract logics $\mathbb{L}$ and $\mathbb{L}'$. By Proposition 1.7 we have $\widetilde{\Omega}(\mathbb{L}) = h^{-1}[\widetilde{\Omega}(\mathbb{L}')]$. Now if $\mathbb{L}'$

has the congruence property and $\langle a, b \rangle \in \Lambda(\mathrm{C})$, it follows that $\langle h(a), h(b) \rangle \in \Lambda(\mathrm{C}') = \widetilde{\Omega}(\mathbb{L}')$ and so $\langle a, b \rangle \in \widetilde{\Omega}(\mathbb{L})$; therefore $\Lambda(\mathrm{C}) = \widetilde{\Omega}(\mathbb{L})$ which means that $\mathbb{L}$ has the congruence property. Conversely, suppose that $\mathbb{L}$ does have it, and that $\langle a', b' \rangle \in \Lambda(\mathrm{C}')$. Let $a' = h(a)$ and $b' = h(b)$ for some $a, b \in A$. Then $h[\mathrm{C}(a)] = \mathrm{C}'(h(a)) = \mathrm{C}'(h(b)) = h[\mathrm{C}(b)]$, and by Proposition 1.5 $\mathrm{C}(a) = h^{-1}[h[\mathrm{C}(a)]] = h^{-1}[h[\mathrm{C}(b)]] = \mathrm{C}(b)$, that is, $\langle a, b \rangle \in \Lambda(\mathrm{C}) = \widetilde{\Omega}(\mathbb{L}) = h^{-1}[\widetilde{\Omega}(\mathbb{L}')]$ which yields $\langle a', b' \rangle \in \widetilde{\Omega}(\mathbb{L}')$. This proves that $\mathbb{L}'$ has the congruence property. $\dashv$

These definitions apply obviously to sentential logics; the Frege relation is then just interderivability, while for any theory Γ, the relation $\Lambda_{\mathcal{S}}(\Gamma)$ is the interderivability relation modulo the theory Γ. With regard to the behaviour of these two relations, we define two kinds of sentential logics of particular interest:

DEFINITION 2.41. *A sentential logic $\mathcal{S}$ is **selfextensional** when, considered as an abstract logic, it has the congruence property, that is, when $\Lambda(\mathcal{S}) = \widetilde{\Omega}(\mathcal{S})$.*

*A sentential logic $\mathcal{S}$ is **strongly selfextensional**[19] when all its full models have the congruence property, that is, when for any $\mathbb{L} \in \mathbf{FMod}\mathcal{S}$, $\widetilde{\Omega}(\mathbb{L}) = \Lambda(\mathbb{L})$.*

The notion of a selfextensional logic has been introduced and studied by Wójcicki (see Section 5.6 of his [1988]). A sentential logic $\mathcal{S}$ is selfextensional if and only if it satisfies the following metalogical property:

If $\varphi_i \dashv\vdash_{\mathcal{S}} \psi_i$ for all $i < n$, then $\varpi \varphi_0 \ldots \varphi_{n-1} \vdash_{\mathcal{S}} \varpi \psi_0 \ldots \psi_{n-1}$

for each basic operation ϖ of the similarity type, where n is the arity of the operation. A sentential logic is strongly selfextensional when this property is inherited, in the obvious sense, by all its full models. In view of Proposition 2.40 and Corollary 2.12, we observe:

PROPOSITION 2.42. *A sentential logic $\mathcal{S}$ is strongly selfextensional iff every abstract logic of the form $\langle \boldsymbol{A}, \mathcal{F}i_{\mathcal{S}} \boldsymbol{A} \rangle$ has the congruence property.* $\dashv$

Thus the congruence property is hard-wired inside strongly selfextensional logics, since by taking all filters on any algebra we always obtain it. It is clear that any strongly selfextensional logic is also selfextensional.

OPEN PROBLEM. *Is every selfextensional logic strongly selfextensional?*

---

[19]This property is defined by requiring that all full models satisfy the congruence property, which defines selfextensionality for sentential logics. Accordingly, in later publications the more descriptive term *fully selfextensional* has been adopted, beginning with Definition 16 in Font [2003b].

In Chapter 4 we answer this affirmatively for two very large classes of logics, those with Conjunction and those with the Deduction Theorem, but a general answer is not known[20].

Now we give two properties showing that sentential logics in these classes have a nice behaviour.

PROPOSITION 2.43. *Let S be any selfextensional sentential logic. Then an equation $\varphi \approx \psi$ is valid in the variety $\mathbf{K}_S$ if and only if $\varphi \dashv\vdash_S \psi$.*

PROOF. By (1.6), an equation $\varphi \approx \psi$ holds in $\mathbf{K}_S$ iff $\gamma(\varphi, \vec{q}) \dashv\vdash_S \gamma(\psi, \vec{q})$ for any $\gamma(p, \vec{q}) \in Fm$. By taking $\gamma(p, \vec{q}) = p$ we obtain one of the implications, which holds in general. Conversely, if $\varphi \dashv\vdash_S \psi$ and S is selfextensional, then the congruence property implies the replacement property, that is, that for any $\gamma(p, \vec{q}) \in Fm$, $\gamma(\varphi, \vec{q}) \dashv\vdash_S \gamma(\psi, \vec{q})$, and this tells us that $\varphi \approx \psi$ holds in $\mathbf{K}_S$. $\dashv$

The next result is the improvement of Theorem 2.36 we announced before.

THEOREM 2.44. *If S is a strongly selfextensional sentential logic, then the category $\mathfrak{L}_+^*$ of all reduced full models of S with all logical morphisms is a full reflective subcategory of the category $\mathfrak{L}_+$ of all full models of S with all logical morphisms; and the reflector is the functor associated with the process of "reduction": $\mathbb{L} \longmapsto \mathbb{L}^*$.*

PROOF. The proof follows the lines of the proof of Theorem 2.36 except for the proof of the central point. $\mathfrak{L}_+^*$ is trivially a full subcategory of $\mathfrak{L}_+$. In order to check that the process of reduction $\mathbb{L} \mapsto \langle \mathbb{L}^*, \pi_{\mathbb{L}} \rangle$ (where $\pi_{\mathbb{L}} : \mathbb{L} \to \mathbb{L}^*$ is the canonical projection) gives the announced reflector, it is enough to check (see Balbes and Dwinger [1974] I.18.2 for instance) that for any logical morphism $f : \mathbb{L} \to \mathbb{L}'$ between an $\mathbb{L} \in \mathbf{FMod}S$ and an $\mathbb{L}' \in \mathbf{FMod}^*S$ there is a unique logical morphism $f^* : \mathbb{L}^* \to \mathbb{L}'$ such that $f^* \circ \pi_{\mathbb{L}} = f$. Since $\pi_{\mathbb{L}}$ is a bilogical morphism, we can use Proposition 1.15 if we prove that $\ker f \supseteq \ker \pi_{\mathbb{L}} = \widetilde{\Omega}(\mathbb{L})$. Let $a, b \in A$ with $\langle a, b \rangle \in \widetilde{\Omega}(\mathbb{L})$; since S is strongly selfextensional, $\mathbb{L}$ has the congruence property, so we have $\langle a, b \rangle \in \Lambda(\mathbb{L})$, that is, for any $T \in \mathcal{C}_{\mathbb{L}}$, $a \in T$ iff $b \in T$. Since $\mathbb{L}'$ is a reduced full model of S, $\mathcal{C}_{\mathbb{L}}' = \mathcal{F}i_S A'$, and since f is a logical morphism, this implies that for any $F \in \mathcal{F}i_S A'$, $f^{-1}[F] \in \mathcal{C}_{\mathbb{L}}$, and so $f(a) \in F$ iff $f(b) \in F$, that is, $\langle f(a), f(b) \rangle \in \Lambda(\mathbb{L}') = \widetilde{\Omega}(\mathbb{L}') = Id_{A'}$ again because $\mathbb{L}'$ has the congruence property and is reduced. Thus $f(a) = f(b)$ which proves $\langle a, b \rangle \in \ker f$. Then by Proposition 1.15 there is a unique logical morphism $f^* : \mathbb{L}^* \to \mathbb{L}'$ with $f^* \circ \pi_{\mathbb{L}} = f$ as was desired. $\dashv$

---

[20]The general question was answered negatively in Babyonyshev [2003] by providing an example of a selfextensional logic that is not strongly selfextensional.

The Property of Conjunction

DEFINITION 2.45. *Let* $\mathbb{L} = \langle A, C \rangle$ *be an abstract logic of some similarity type, and let* $\wedge$ *be a binary operation symbol, either primitive or defined by a term. We say that* $\mathbb{L}$ *has the **Property of Conjunction (PC)** with respect to* $\wedge$ *when for any* $a, b \in A$,

$$C(a, b) = C(a \wedge b). \tag{PC}$$

In the literature it is also said that an abstract logic $\mathbb{L}$ is *conjunctive* or that the binary term $\wedge$ is a *Conjunction for* $\mathbb{L}$ when $\mathbb{L}$ has the PC with respect to $\wedge$. Normally we will omit the reference "with respect to $\wedge$" since the operation involved will be clear from context. The following observations are straightforward and/or well-known:

1. $\mathbb{L}$ has the PC iff for any $T \in C$ and any $a, b \in A$, $a \wedge b \in T$ iff $a \in T$ and $b \in T$.
2. The Property of Conjunction is preserved under bilogical morphisms (see Font and Verdú [1991], Proposition 4.1). In particular, $\mathbb{L}$ has the PC iff $\mathbb{L}^*$ has the PC.
3. If $\mathbb{L}$ has the PC with respect to $\wedge$ then $\Lambda(\mathbb{L})$ is a congruence with respect to $\wedge$, and for every $F \subseteq A$, $\Lambda_{\mathbb{L}}(F)$ is also a congruence with respect to $\wedge$.
4. A sentential logic $\mathcal{S}$ has the PC iff the following rules hold for $\mathcal{S}$:

$$\varphi \wedge \psi \vdash \psi \ , \ \ \varphi \wedge \psi \vdash \varphi \ \ \text{and} \ \ \{\varphi, \psi\} \vdash \varphi \wedge \psi.$$

5. If a sentential logic $\mathcal{S}$ has the PC then all its models also have the PC (with respect to the same operation). In particular, all its full models have the PC.

Moreover we can prove:

PROPOSITION 2.46. *Let* $\mathcal{S}$ *be a sentential logic with the PC. Then every finitary model of* $\mathcal{S}$ *(having no theorems if* $\mathcal{S}$ *does not) which satisfies the congruence property is a full model of* $\mathcal{S}$.

PROOF. Suppose that $\mathbb{L}$ is a finitary model for $\mathcal{S}$, that is, $\mathcal{C} \subseteq \mathcal{F}i_{\mathcal{S}}A$, such that $\emptyset \in \mathcal{C}$ iff $\mathcal{S}$ does not have theorems, and with the congruence property. We must prove that $\mathcal{C}^* = \mathcal{F}i_{\mathcal{S}}A^*$. If $F \in \mathcal{C}^*$ then also $F \in \mathcal{F}i_{\mathcal{S}}A^*$ by Proposition 1.19, since $\pi^{-1}[F] \in \mathcal{C}$. Conversely, let $F \in \mathcal{F}i_{\mathcal{S}}A^*$. If $F = \emptyset$ then $\mathcal{S}$ cannot have theorems, and by assumption $\emptyset \in \mathcal{C}$ so also $\emptyset \in \mathcal{C}^*$. If $F \neq \emptyset$ then, by finitarity of $\mathbb{L}^*$, for any $a \in C^*(F)$ there are $a_1, \ldots, a_n \in F$ such that $a \in C^*(a_1, \ldots, a_n)$. But $\mathbb{L}$ has the PC because it is a model of $\mathcal{S}$, so $\mathbb{L}^*$ also has it, therefore $a \in C^*\big(a_1 \wedge (\ldots \wedge a_n)\big)$ and this implies $C^*\big(a \wedge (a_1 \wedge (\ldots \wedge a_n))\big) = C^*\big(a_1 \wedge (\ldots \wedge a_n)\big)$. But since $\mathbb{L}$ has the congruence property, by 2.40 $\mathbb{L}^*$ also has it, and since it is reduced, we conclude that $a \wedge \big(a_1 \wedge (\ldots \wedge a_n)\big) = a_1 \wedge (\ldots \wedge a_n)$. Now F is an $\mathcal{S}$-filter and $\mathcal{S}$ has the PC; this implies that $a_1 \wedge (\ldots \wedge a_n) \in F$, therefore also $a \wedge \big(a_1 \wedge (\ldots \wedge a_n)\big) \in F$, and from this it follows that $a \in F$. This proves that $C^*(F) = F$, that is, $F \in \mathcal{C}^*$. This completes the proof that $\mathcal{C}^* = \mathcal{F}i_{\mathcal{S}} A^*$, and so $\mathbb{L}$ is a full model of $\mathcal{S}$. $\qquad\dashv$

In Section 4.2 we will prove the converse of this result for selfextensional logics: every full model of a selfextensional logic with the PC has the congruence property, and therefore such a logic is strongly selfextensional. Thus we see that the PC is a very strong property: it makes the congruence property (for *all* the connectives of the language) to be inherited from the logic by all its full models.

The Deduction-Detachment Theorem

We will consider here only the more classical version of the Deduction Theorem, that is, the one concerning only a binary connective, either primitive or defined by a single term; more general versions, including weaker "Deduction Theorems", have been dealt with in Blok and Pigozzi [1991], [1989b], Czelakowski [1985], [1986] and Czelakowski and Dziobiak [1991].

Strictly speaking, the name of *Deduction Theorem* is usually applied to just the implication

$$\Gamma, \varphi \vdash_{\mathcal{S}} \psi \quad \Longrightarrow \quad \Gamma \vdash_{\mathcal{S}} \varphi \to \psi, \tag{DT}$$

while the converse one receives the name of *Modus Ponens* (MP) or *Detachment*; we will follow this distinction, since the metalogical status of both properties is very different: while the MP is equivalent to a Hilbert-style rule, and so is inherited by all models of a sentential logic, this is not the case of the DT; the latter is, however, inherited by all full models.

DEFINITION 2.47. *Let $\to$ be a binary operation symbol, either primitive or defined by a term, and let $\mathbb{L} = \langle A, C \rangle$ be an abstract logic. We say that $\mathbb{L}$ satisfies, with respect to $\to$, the:*

(1) *Modus Ponens (MP) when for any $a, b \in A$ and any $X \subseteq A$,*

$$a \to b \in C(X) \quad \text{implies} \quad b \in C(X, a). \tag{MP}$$

(2) *Deduction Theorem (DT) when for any $a, b \in A$ and any $X \subseteq A$,*

$$b \in C(X, a) \quad \text{implies} \quad a \to b \in C(X). \tag{DT}$$

(3) *Deduction-Detachment Theorem (DDT) when it satisfies the MP and the DT.*

We will usually omit the reference "with respect to $\rightarrow$" since only one such operation will be considered. The following observations are straightforward or well-known:

1. If an abstract logic $\mathbb{L} = \langle A, \mathrm{C} \rangle$ satisfies the DDT then it has theorems, namely for any $a \in A$, $\mathrm{C}(a \rightarrow a) = \mathrm{C}(\emptyset)$. Some particular theorems of such abstract logics will be used, specially in the case of sentential logics; we highlight the following, for all $a, b, c \in A$:

$$a \rightarrow a$$
$$a \rightarrow (b \rightarrow a)$$
$$(a \rightarrow (b \rightarrow c)) \rightarrow ((a \rightarrow b) \rightarrow (a \rightarrow c))$$

2. As a consequence of 1, if a sentential logic $\mathcal{S}$ has the DDT then every $\mathcal{S}$-filter is non-empty, and thus every model of it, as well as every full model, has theorems.

3. An abstract logic $\mathbb{L}$ has the MP iff for every $a, b \in A$, $b \in \mathrm{C}(a, a \rightarrow b)$, and also iff for every closed set $T \in \mathcal{C}$, if $a \in T$ and $a \rightarrow b \in T$ then $b \in T$; informally we refer to this property as being *closed under the MP*.

4. A sentential logic $\mathcal{S}$ has the MP if and only if the following is a rule of $\mathcal{S}$: $\{\varphi, \varphi \rightarrow \psi\} \vdash_{\mathcal{S}} \psi$. As a consequence, each $\mathcal{S}$-filter is closed under the MP and every model of $\mathcal{S}$ (and hence every full model) also has the MP.

5. The DDT is preserved under bilogical morphisms (see the Corollary to Proposition 6 of Verdú [1987]). In particular, $\mathbb{L}$ has the DDT iff $\mathbb{L}^{*}$ has the DDT. Actually this holds separately for the MP and for the DT.

6. If $\mathbb{L}$ has the DDT then for any $F \subseteq A$, the Frege relation $\Lambda_{\mathbb{L}}(F)$ is a congruence with respect to $\rightarrow$.

Thus the MP is inherited by all models of a sentential logic satisfying it. Next we see that the DT (and hence the DDT) is inherited by full models, a fact that is essentially contained in Theorem 2.2 of Czelakowski [1985].

THEOREM 2.48. *If $\mathcal{S}$ has the DDT then every full model of $\mathcal{S}$ has the DDT.*

PROOF. Assume that $\mathcal{S}$ has the DDT, that is, the MP and the DT. By Corollary 2.12, it will be enough to prove that every abstract logic of the form $\langle A, \mathcal{F}i_{\mathcal{S}} A \rangle$ has the DDT. As we have already noticed, every $\mathcal{S}$-filter is closed under the MP, thus $\langle A, \mathcal{F}i_{\mathcal{S}} A \rangle$ has the MP. Now we have to prove that for all $X \subseteq A$ and all $a, b \in A$, if $b \in \mathrm{Fi}_{\mathcal{S}}^{A}(X, a)$ then $a \rightarrow b \in \mathrm{Fi}_{\mathcal{S}}^{A}(X)$. We use the characterization of $\mathrm{Fi}_{\mathcal{S}}^{A}(X, a)$ given in Lemma 1.18: $\mathrm{Fi}_{\mathcal{S}}^{A}(X, a) = \bigcup \{X_n : n \in \omega\}$, where the X_n are defined as in 1.18, starting with $X_0 = X \cup \{a\}$. Then we prove by induction on n that if $b \in X_n$ then $a \rightarrow b \in \mathrm{Fi}_{\mathcal{S}}^{A}(X)$: Assume that $n = 0$ and $b \in X_0 =$

$X \cup \{a\}$; if $b \in X$ then since $b \to (a \to b) \in \operatorname{Fi}_{\mathcal{S}}^{\mathbf{A}}(X)$ also $a \to b \in \operatorname{Fi}_{\mathcal{S}}^{\mathbf{A}}(X)$, and if $b = a$ then $a \to b = a \to a \in \operatorname{Fi}_{\mathcal{S}}^{\mathbf{A}}(X)$. Now assume $n \geqslant 1$ and the thesis true for n, and let $b \in X_{n+1}$: there are a finite $\Gamma \subseteq Fm$ and $\varphi \in Fm$ such that $\Gamma \vdash_{\mathcal{S}} \varphi$ and there is $h \in \operatorname{Hom}(\boldsymbol{Fm}, \boldsymbol{A})$ such that $h[\Gamma] \subseteq X_n$ and $h(\varphi) = b$. If $\Gamma = \emptyset$ then trivially $b \in \operatorname{Fi}_{\mathcal{S}}^{\mathbf{A}}(X)$. Now assume $\Gamma = \{\psi_1, \ldots, \psi_k\}$. Let $q \in Var$ be any variable not appearing in $\psi_1, \ldots \psi_k, \varphi$; using the DDT and its consequences for $\vdash_{\mathcal{S}}$ we obtain $\{q \to \psi_i : i = 1, \ldots, k\} \vdash_{\mathcal{S}} q \to \varphi$. Define $h' \in \operatorname{Hom}(\boldsymbol{Fm}, \boldsymbol{A})$ such that $h'(p) = h(p)$ if $p \neq q$ while $h'(q) = a$. By the inductive hypothesis $h'(q) \to h'(\psi_i) = a \to h(\psi_i) \in \operatorname{Fi}_{\mathcal{S}}^{\mathbf{A}}(X)$ for $i = 1, \ldots, k$, therefore $a \to b = a \to h(\varphi) = h'(q) \to h'(\varphi) \in \operatorname{Fi}_{\mathcal{S}}^{\mathbf{A}}(X)$. This finishes the inductive proof. Therefore $a \to b \in \operatorname{Fi}_{\mathcal{S}}^{\mathbf{A}}(X)$.   ⊣

In Section 4.3 we will use the DT to find a characterization of full models among the class of all (finitary) models; in the meantime we can prove a partial result we will need there:

PROPOSITION 2.49. *Let $\mathcal{S}$ be a sentential logic with the DDT. If $\mathbb{L} = \langle \boldsymbol{A}, \mathrm{C} \rangle$ is a finitary model of $\mathcal{S}$ with the DT and with the congruence property, then $\mathbb{L}$ is a full model of $\mathcal{S}$.*

PROOF. Suppose that $\mathbb{L}$ is a finitary model of $\mathcal{S}$ with the DT and the congruence property. We have that $\mathcal{C} \subseteq \mathcal{Fi}_{\mathcal{S}} \boldsymbol{A}$, and we must prove that $\mathcal{C}^* = \mathcal{Fi}_{\mathcal{S}} \boldsymbol{A}^*$. If $F \in \mathcal{C}^*$ then also $F \in \mathcal{Fi}_{\mathcal{S}} \boldsymbol{A}^*$ by Proposition 1.19, since $\pi^{-1}[F] \in \mathcal{C}$. Conversely, let $F \in \mathcal{Fi}_{\mathcal{S}} \boldsymbol{A}^*$. Since $\mathcal{S}$ has the DDT, $F \neq \emptyset$, so by finitarity of $\mathbb{L}^*$, for any $a \in \mathrm{C}^*(F)$ there are $a_1, \ldots, a_n \in F$ such that $a \in \mathrm{C}^*(a_1, \ldots, a_n)$. But $\mathbb{L}$ has the DDT by assumption, so $\mathbb{L}^*$ also has it, therefore $a_1 \to (\ldots (a_n \to a) \ldots) \in \mathrm{C}^*(\emptyset) = \mathrm{C}^*(a \to a)$ and this implies $\mathrm{C}^*\big(a_1 \to (\ldots (a_n \to a) \ldots)\big) = \mathrm{C}^*(a \to a)$. But since $\mathbb{L}$ has the congruence property, by 2.40 $\mathbb{L}^*$ also has it, and since it is reduced, we conclude that $a_1 \to (\ldots (a_n \to a) \ldots) = a \to a \in F$. Since F is an $\mathcal{S}$-filter and $\mathcal{S}$ has the MP, this implies that $a \in F$. This proves that $\mathrm{C}^*(F) = F$, that is, $F \in \mathcal{C}^*$. This completes the proof that $\mathcal{C}^* = \mathcal{Fi}_{\mathcal{S}} \boldsymbol{A}^*$, that is, $\mathbb{L}$ is a full model of $\mathcal{S}$.   ⊣

As we will prove in Corollary 4.30, if $\mathcal{S}$ is selfextensional then the converse of this property also holds.

The Property of Disjunction

This property, which should not be confused with the so-called "Disjunction Property" of some intermediate logics (stating that if $\vdash_{\mathcal{S}} \varphi \vee \psi$ then $\vdash_{\mathcal{S}} \varphi$ or $\vdash_{\mathcal{S}} \psi$), corresponds to the method of *Proof by Cases* of traditional logic; in the literature it is also said that a logic $\mathbb{L}$ is *disjunctive* when it satisfies this property, see Czelakowski [1984]:

DEFINITION 2.50. *An abstract logic* $\mathbb{L} = \langle A, C \rangle$ *satisfies the* **Property of Disjunction (PDI)** *with respect to a binary operation symbol* $\vee$, *either primitive or defined by a term, when, for any* $X \subseteq A$, $a, b \in A$,

$$C(X, a \vee b) = C(X, a) \cap C(X, b). \tag{PDI}$$

Some easy or well-known consequences are:

1. A sentential logic S satisfies the PDI iff the following rules hold: The two Hilbert-style rules: $\varphi \vdash_S \varphi \vee \psi$, $\varphi \vdash_S \psi \vee \varphi$ and the Gentzen-style rule:

$$\frac{\Gamma, \psi_1 \vdash \varphi \qquad \Gamma, \psi_2 \vdash \varphi}{\Gamma, \psi_1 \vee \psi_2 \vdash \varphi}.$$

2. If a sentential logic S satisfies the PDI then the following Hilbert-style rules also hold: $\varphi \vee \psi \dashv\vdash_S \psi \vee \varphi$ and $\varphi \dashv\vdash_S \varphi \vee \varphi$.
3. The PDI is preserved under bilogical morphisms. In particular, $\mathbb{L}$ satisfies the PDI iff $\mathbb{L}^*$ satisfies it. See Font and Verdú [1991], Proposition 4.1.
4. If $\mathbb{L} = \langle A, C \rangle$ satisfies the PDI then an easy inductive argument shows that for any $a_1, \ldots, a_n, b \in A$ and any $X \subseteq A$,

$$C(X, a_1 \vee b, \ldots, a_n \vee b) = C(X, a_1, \ldots, a_n) \cap C(X, b).$$

LEMMA 2.51. *Let* S *be a sentential logic satisfying the PDI and assume that* $\psi_1, \ldots, \psi_n \vdash_S \varphi$. *Then for any* ξ, $\psi_1 \vee \xi, \ldots, \psi_n \vee \xi \vdash_S \varphi \vee \xi$.

PROOF. From the generalization of the PDI mentioned in item 4 above we can obtain, as a particular case, that for any $\psi_1, \ldots, \psi_n, \xi \in Fm$, $\mathrm{Cn}_S(\psi_1 \vee \xi, \ldots, \psi_n \vee \xi) = \mathrm{Cn}_S(\psi_1, \ldots, \psi_n) \cap \mathrm{Cn}_S(\xi)$. Now, $\varphi \in \mathrm{Cn}_S(\psi_1, \ldots, \psi_n)$ by assumption, and obviously $\xi \in \mathrm{Cn}_S(\xi)$. But the PDI implies that $\varphi \vee \xi \in \mathrm{Cn}_S(\varphi) \cap \mathrm{Cn}_S(\xi)$. Therefore we obtain $\varphi \vee \xi \in \mathrm{Cn}_S(\psi_1 \vee \xi, \ldots, \psi_n \vee \xi)$ as desired. $\dashv$

Next we see that the PDI is inherited by full models; the essential part of the proof is also mentioned in Czelakowski [1984].

THEOREM 2.52. *If* S *is a sentential logic with the PDI then every full model of* S *satisfies the PDI as well.*

PROOF. By Corollary 2.12 it will be enough to prove that, for any A and any $X \cup \{a, b\} \subseteq A$, $\mathrm{Fi}_S^A(X, a \vee b) = \mathrm{Fi}_S^A(X, a) \cap \mathrm{Fi}_S^A(X, b)$. From the Hilbert-style rules mentioned in item 2 above it follows that $\mathrm{Fi}_S^A(X, a \vee b) \subseteq \mathrm{Fi}_S^A(X, a) \cap$

$\mathrm{Fi}_\mathcal{S}^{\boldsymbol{A}}(X, b)$. In order to establish the reverse inclusion we first prove that for any $a, b, c \in A$,

$$c \in \mathrm{Fi}_\mathcal{S}^{\boldsymbol{A}}(X, a) \text{ implies } c \vee b \in \mathrm{Fi}_\mathcal{S}^{\boldsymbol{A}}(X, a \vee b). \tag{$*$}$$

For this consider the characterization $\mathrm{Fi}_\mathcal{S}^{\boldsymbol{A}}(X, a) = \bigcup \{X_n : n \in \omega\}$ of Lemma 1.18. Let us prove by induction on n that if $c \in X_n$ then $c \vee b \in \mathrm{Fi}_\mathcal{S}^{\boldsymbol{A}}(X, a \vee b)$. Since $X_0 = X \cup \{a\}$, the case $n = 0$ is trivial. Assuming the property is true for n, let $c \in X_{n+1}$: this means that there are $\varphi, \psi_1, \ldots \psi_k \in Fm$ such that $\psi_1, \ldots \psi_k \vdash_\mathcal{S} \varphi$ and there is $h \in \mathrm{Hom}(\boldsymbol{Fm}, \boldsymbol{A})$ with $h(\psi_i) \in X_n$ and $h(\varphi) = c$. Now choose some variable q not appearing in these formulas, and modify h at q in order to obtain $h' \in \mathrm{Hom}(\boldsymbol{Fm}, \boldsymbol{A})$ such that $h'(q) = b$ and $h'(\psi_i) = h(\psi_i)$. By the induction hypothesis $h'(\psi_i \vee q) = h'(\psi_i) \vee b \in \mathrm{Fi}_\mathcal{S}^{\boldsymbol{A}}(X, a \vee b)$, and since by Lemma 2.51 $\psi_1 \vee q, \ldots \psi_k \vee q \vdash_\mathcal{S} \varphi \vee q$, it follows that $c \vee b = h'(\varphi \vee q) \in \mathrm{Fi}_\mathcal{S}^{\boldsymbol{A}}(X, a \vee b)$. Thus $(*)$ is proved and using it we can now prove the remaining part of the PDI: Take any $c \in \mathrm{Fi}_\mathcal{S}^{\boldsymbol{A}}(X, a) \cap \mathrm{Fi}_\mathcal{S}^{\boldsymbol{A}}(X, b)$. From $c \in \mathrm{Fi}_\mathcal{S}^{\boldsymbol{A}}(X, a)$ it follows $c \vee b \in \mathrm{Fi}_\mathcal{S}^{\boldsymbol{A}}(X, a \vee b)$, and from $c \in \mathrm{Fi}_\mathcal{S}^{\boldsymbol{A}}(X, b)$ it follows $c \vee c \in \mathrm{Fi}_\mathcal{S}^{\boldsymbol{A}}(X, b \vee c)$. Since $c \in \mathrm{Fi}_\mathcal{S}^{\boldsymbol{A}}(c \vee c)$ and $b \vee c \in \mathrm{Fi}_\mathcal{S}^{\boldsymbol{A}}(c \vee b)$ we conclude that $c \in \mathrm{Fi}_\mathcal{S}^{\boldsymbol{A}}(X, a \vee b)$, as had to be proved. $\dashv$

The fact that not every model of $\mathcal{S}$ inherits the PDI is shown in Section 5.1.1 by a simple example. The Property of Disjunction can be generalized by using a finite set of terms instead of a single term.

The two forms of Reductio ad Absurdum

Now we consider the forms of *Reductio ad Absurdum* that hold in Intuitionistic Logic and in Classical Logic:

DEFINITION 2.53. *Let $\neg$ be a unary operation symbol, either primitive or defined by a term. An abstract logic $\mathbb{L} = \langle \boldsymbol{A}, \mathrm{C} \rangle$ satisfies the **Property of Intuitionistic Reductio ad Absurdum (PIRA)** with respect to $\neg$ when for any $X \subseteq A$ and any $a \in A$,*

$$\neg a \in \mathrm{C}(X) \iff \mathrm{C}(X, a) = A;$$

*and it satisfies the **Property of Reductio ad Absurdum (PRA)** with respect to $\neg$ when for any $X \subseteq A$ and any $a \in A$,*

$$a \in \mathrm{C}(X) \iff \mathrm{C}(X, \neg a) = A.$$

It is easy to see that an abstract logic satisfies the PRA if and only if it satisfies both the PIRA and that $a \in C(\neg\neg a)$. Speaking of sentential logics, this last property is a Hilbert-style rule, which is inherited by all models; hence the problem of inheritance of the PRA by full models reduces to that of the PIRA.

The PIRA is not inherited in general by all full models of a sentential logic having it: Take as an example the $\neg$-fragment $\mathcal{S}$ of intuitionistic logic: In Porębska and Wroński [1975] it is proved that this fragment is characterized precisely by the PIRA (it is the weakest sentential logic having it, when the language has just negation), and it does not have theorems. Now every one-element set $A = \{a\}$ provides us with a counterexample, since we must have $\neg a = a$: Clearly $\mathcal{F}i_{\mathcal{S}} A = \{\emptyset, A\}$ and the abstract logic $\langle A, \mathcal{F}i_{\mathcal{S}} A \rangle$, which is a full model of $\mathcal{S}$, does not satisfy the PIRA: $\mathrm{Fi}_{\mathcal{S}}^{A}(a) = A$ but $\neg a \notin \mathrm{Fi}_{\mathcal{S}}^{A}(\emptyset) = \emptyset$.

The difficulties revealed by the analysis of the general case of this problem tell us that negation is a difficult connective to deal with alone. But one of the main results of Section 4.2 will enable us to prove that if $\mathcal{S}$ is a selfextensional logic with the PC and the PIRA then every full model of $\mathcal{S}$ has the PC and the PIRA; and in the case where conjunction and negation are the only connectives of the language we will be able to remove the assumption of selfextensionality; see Propositions 4.34 and 4.35. At this moment we can treat the case of the DDT and the PIRA together. Actually, in the presence of the DDT, the PIRA is equivalent to a very simple requirement.

If $\mathbb{L} = \langle A, C \rangle$ is an abstract logic, we say that an element $\bot \in A$ is an **inconsistent element** when $C(\bot) = A$; authors in the field of *paraconsistent logics* sometimes prefer to call such elements *trivial*. Then:

LEMMA 2.54. *Let $\mathbb{L} = \langle A, C \rangle$ be an abstract logic with the DDT with respect to a binary operation symbol $\rightarrow$. Then $\mathbb{L}$ satisfies the PIRA with respect to some unary operation symbol $\neg$ if and only if $\mathbb{L}$ has an inconsistent element $\bot$. Moreover, in this situation, $C(\neg a) = C(a \rightarrow \bot)$ for any $a \in A$.*

PROOF. It is trivial to check (using the DDT) that, if $\mathbb{L}$ satisfies the PIRA with respect to $\neg$ then for any $a \in A$ the element $\neg(a \rightarrow a)$ is inconsistent, and that if $\bot$ is an inconsistent element, then $\mathbb{L}$ satisfies the PIRA with respect to the operation $\neg a = a \rightarrow \bot$. In general, if $\bot$ is inconsistent, from the MP it follows that $C(a, a \rightarrow \bot) = A$, and therefore by the PIRA $\neg a \in C(a \rightarrow \bot)$; since $\neg a \in C(\neg a)$, we have that $\bot \in C(a, \neg a)$, and by the DDT this implies $a \rightarrow \bot \in C(\neg a)$; therefore we have shown that $C(\neg a) = C(a \rightarrow \bot)$. $\dashv$

Since having an inconsistent element is a property clearly inherited by any model, it follows from Theorem 2.48 and the previous lemma:

COROLLARY 2.55. *If a sentential logic satisfies the DDT and the PIRA then all its full models satisfy them.*   ⊣

Some rules of introduction of modality

One of the strongest metalogical properties of *normal modal logics* is the so-called **Rule of Necessitation**. In its *strong form* it is:

$$\varphi \vdash \Box\varphi,$$

and like all Hilbert-style rules, it is inherited by every model; so it is not especially interesting to consider it here. The same rule has a *weak form*, which has also been considered in the literature:

$$\text{If } \vdash \varphi \text{ then } \vdash \Box\varphi.$$

However, this Gentzen-style rule is but a particular case of the rule more commonly taken in many Gentzen-style formulations of systems of modal logic as a rule for introduction of the necessity operator, see Zeman [1973],

$$(\mathrm{I}\Box) \quad \frac{\Gamma \vdash \varphi}{\Box\Gamma \vdash \Box\varphi},$$

where $\Box\Gamma = \{\Box\gamma : \gamma \in \Gamma\}$. Actually, the same rule holds for the possibility operator $\Diamond$ in the place of $\Box$, and also for a number of other unary operators of modal character (temporal, dynamic, etc.), and even for double negation $\neg\neg$, which in some logics has been shown to have a modal behaviour, see Došen [1986]. Accordingly, let $\#$ be an arbitrary unary operation symbol, either primitive or defined by a term; we say that an abstract logic $\mathbb{L} = \langle A, \mathrm{C} \rangle$ is *closed under introduction of $\#$* when for any $X \subseteq A$, $\#\mathrm{C}(X) \subseteq \mathrm{C}(\#X)$, that is, when $a \in \mathrm{C}(X)$ implies $\#a \in \mathrm{C}(\#X)$. Then:

PROPOSITION 2.56. *If $\mathcal{S}$ is a sentential logic closed under introduction of a unary connective $\#$ then all its full models are also closed under introduction of the same connective.*

PROOF. In Jansana [1995] it has been proved that the property of being closed under introduction of a unary connective is preserved under bilogical morphisms. Therefore, as usual, it will be enough to prove, for any A, any $X \subseteq A$ and any $a \in A$, that if $a \in \mathrm{Fi}_{\mathcal{S}}^{A}(X)$ then $\#a \in \mathrm{Fi}_{\mathcal{S}}^{A}(\#X)$. Put $\mathrm{Fi}_{\mathcal{S}}^{A}(X) = \bigcup\{X_n : n \in \omega\}$ as in Proposition 1.18, and prove by induction that if $a \in X_n$ then $\#a \in \mathrm{Fi}_{\mathcal{S}}^{A}(\#X)$. Since $X_0 = X$, the case $n = 0$ is trivial. If $a \in X_{n+1}$ then for some formulas $\psi_1, \ldots, \psi_k \vdash_{\mathcal{S}} \varphi$ and there is an homomorphism h such that $h(\psi_i) \in X_n$ and $h(\varphi) = a$. By induction $h(\#\psi_i) = \#h(\psi_i) \in \mathrm{Fi}_{\mathcal{S}}^{A}(\#X)$, and by introduction of $\#$ for $\mathcal{S}$ we have $\#\psi_1, \ldots, \#\psi_k \vdash_{\mathcal{S}} \#\varphi$, therefore $\#a = \#h(\varphi) = h(\#\varphi) \in$

$\mathrm{Fi}_{\mathcal{S}}^{\boldsymbol{A}}(X)$. Therefore this holds for every n, and thus $\#\mathrm{Fi}_{\mathcal{S}}^{\boldsymbol{A}}(X) \subseteq \mathrm{Fi}_{\mathcal{S}}^{\boldsymbol{A}}(\#X)$, that is, the abstract logic $\langle \boldsymbol{A}, \mathrm{Fi}_{\mathcal{S}}^{\boldsymbol{A}} \rangle$ is closed under introduction of $\#$. $\dashv$

CHAPTER 3

APPLICATIONS TO PROTOALGEBRAIC
AND ALGEBRAIZABLE LOGICS

One of the most important classes of sentential logics from the point of view of their algebraization is the class of the protoalgebraic logics. As defined in Blok and Pigozzi [1986], a sentential logic is **protoalgebraic** when for any $\Gamma \in Th\mathcal{S}$, any two formulas equivalent modulo $\Omega_{Fm}(\Gamma)$ are also $\mathcal{S}$-interderivable modulo Γ; that is, when for any $\Gamma \in Th\mathcal{S}$ and any $\varphi, \psi \in Fm$,

$$\text{if } \langle \varphi, \psi \rangle \in \Omega_{Fm}(\Gamma) \text{ then } \Gamma, \varphi \vdash_{\mathcal{S}} \psi \text{ and } \Gamma, \psi \vdash_{\mathcal{S}} \varphi,$$

or, in our notation, when for any $\Gamma \in Th\mathcal{S}$, $\Omega_{Fm}(\Gamma) \subseteq \Lambda_{\mathcal{S}}(\Gamma)$.

This class of logics was defined and thoroughly studied in Blok and Pigozzi [1986]. Independently, it was considered in Czelakowski [1985], with a different definition and under the name of *non-pathological logics*; the equivalence of the two definitions was proved in Blok and Pigozzi [1992]. From the results in these and subsequent works (such as Blok and Pigozzi [1991], Czelakowski [2001a] and Czelakowski and Dziobiak [1991]) one can reach the conclusion that these logics are precisely the ones whose matrix semantics is particularly well-behaved from the point of view of universal algebra. Among several interesting characterizations of this notion, let us mention that a logic $\mathcal{S}$ is protoalgebraic iff the Leibniz operator Ω_{Fm} on $Th\mathcal{S}$ is monotone with respect to $\subseteq$. This is also equivalent to saying that for any algebra A, the operator Ω_A is monotone on $\mathcal{F}i_{\mathcal{S}}A$ (see Blok and Pigozzi [1986] Theorem 2.4); this property is called the *Compatibility Property*. Let us look more closely into what this property says: Being monotone means that for any A and any $F, G \in \mathcal{F}i_{\mathcal{S}}A$, if $F \subseteq G$ then $\Omega_A(F) \subseteq \Omega_A(G)$. Observe that $\Omega_A(F) \subseteq \Omega_A(G)$ is equivalent to saying that $\Omega_A(F)$ is compatible with G, that is, that G is a union of equivalence classes modulo $\Omega_A(F)$; if we consider the canonical projection $\pi : A \to A/\Omega_A(F)$, another way of expressing the compatibility property is to say that $G = \pi^{-1}\big[\pi[G]\big]$ for all $G \in \mathcal{F}i_{\mathcal{S}}A$ such that $F \subseteq G$. Taking Proposition 1.19 into account, we see that then $\pi[G] \in \mathcal{F}i_{\mathcal{S}}(A/\Omega_A(F))$ and moreover the correspondence $G \mapsto \pi[G]$ establishes a lattice isomorphism between the lattices $(\mathcal{F}i_{\mathcal{S}}A)^F$ and $(\mathcal{F}i_{\mathcal{S}}(A/\Omega_A(F)))^{\pi[F]}$. This fact, a special case of the so-called *Correspondence Theorem* of Blok and Pigozzi [1986], will be used later on in this chapter. Also note that Ω_A is monotone if and only if it commutes with arbitrary intersections, that is, if and only if $\Omega_A(\bigcap\{F_i : i \in I\}) = \bigcap\{\Omega_A(F_i) : i \in I\}$ for any family $\{F_i : i \in I\} \subseteq \mathcal{F}i_{\mathcal{S}}A$.

An important subclass of protoalgebraic logics is that of **algebraizable logics**, introduced in Blok and Pigozzi [1989a]; in this monograph several characterizations are given for this notion, from different points of view. In the present chapter we will establish some properties of algebraizable logics concerning the notions we have introduced in the preceding chapter. Instead of the definition of algebraizable logic, it will be enough for the reader to know Theorem 13.15 of Blok and Pigozzi [1992], which says that a sentential logic $\mathcal{S}$ is algebraizable iff for every algebra A, the Leibniz operator Ω_A is monotone, injective and continuous on $\mathcal{F}i_{\mathcal{S}}A$; **continuity** means that for any upwards directed family $\{F_i : i \in I\} \subseteq \mathcal{F}i_{\mathcal{S}}A$ it holds that $\Omega_A(\bigcup\{F_i : i \in I\}) = \bigcup\{\Omega_A(F_i) : i \in I\}$. With each algebraizable logic $\mathcal{S}$ one can associate a unique quasivariety $\mathbf{K}$, called the **equivalent quasivariety semantics of $\mathcal{S}$**, having several very close relationships with $\mathcal{S}$; one of them is that there are two elementary definable and structural translations between (sets of) formulas and (sets of) equations in such a way that the consequence $\vdash_{\mathcal{S}}$ of the logic becomes equivalent to the equational consequence $\models_{\mathbf{K}}$ associated with the class $\mathbf{K}$ (see Definition 4.13). Another characterization, of special interest here, is that for any algebra A, the Leibniz operator Ω_A is an isomorphism between the lattices $\mathcal{F}i_{\mathcal{S}}A$ and $\mathrm{Con}_{\mathbf{K}}A$; a logic having this property relative to a quasivariety $\mathbf{K}$ must be algebraizable, and the class $\mathbf{K}$ is its equivalent quasivariety semantics. We will see in this chapter that a non-algebraizable logic can also have this property relative to a class $\mathbf{K}$, but it will not be a quasivariety. And we will relate this class $\mathbf{K}$ with the class $\mathbf{Alg}^*\mathcal{S}$ and with the class $\mathbf{Alg}\mathcal{S}$.

Since we always have that $\Omega_A(\emptyset) = A \times A$, it follows from the definition that the only protoalgebraic logic without theorems is the one satisfying $\varphi \vdash_{\mathcal{S}} \psi$ for all $\varphi, \psi \in Fm$, that is, the logic characterized by $Th\mathcal{S} = \{\emptyset, Fm\}$; this logic is called **almost inconsistent** in Czelakowski [2001a], and appears as a counterexample or as the only pathological case in a variety of situations. The compatibility property also yields the following characterizations of protoalgebraic logics, which use the Tarski congruence:

PROPOSITION 3.1. *For any sentential logic S the following conditions are equivalent:*

(i) S *is protoalgebraic.*

(ii) *For any A and any closure system $C \subseteq \mathcal{F}i_S A$, $\tilde{\Omega}_A(C) = \Omega_A(C(\emptyset))$.*

(iii) *For any A and any $F \in \mathcal{F}i_S A$, $\tilde{\Omega}_A((\mathcal{F}i_S A)^F) = \Omega_A(F)$.*

(iv) *For any $\Gamma \in ThS$, $\tilde{\Omega}(S^\Gamma) = \Omega_{Fm}(\Gamma)$.*

PROOF. (i)$\Rightarrow$(ii) Since $C \subseteq \mathcal{F}i_S A$, the compatibility property implies that Ω_A is also order-preserving on C; then, using this and 1.2 we have

$$\tilde{\Omega}_A(C) = \bigcap \{\Omega_A(T) : T \in C\} = \Omega_A\left(\bigcap \{T : T \in C\}\right) = \Omega_A(C(\emptyset)).$$

(iii) is a particular case of (ii), and (iv) is a particular case of (iii).

(iv)$\Rightarrow$(i) Let $\Gamma, \Gamma' \in ThS$ with $\Gamma \subseteq \Gamma'$. This implies that $\Gamma' \in ThS^\Gamma$ and thus by 1.2, $\tilde{\Omega}(S^\Gamma) \subseteq \Omega_{Fm}(\Gamma')$. Then the assumption gives $\Omega_{Fm}(\Gamma) \subseteq \Omega_{Fm}(\Gamma')$, that is, Ω_{Fm} is order-preserving on ThS, which proves S is protoalgebraic. $\dashv$

In particular, observe that if for any algebra A we denote the least S-filter on A by F_0, then if S is protoalgebraic it satisfies that $\tilde{\Omega}_A(\mathcal{F}i_S A) = \Omega_A(F_0)$. As a consequence, we obtain:

PROPOSITION 3.2. *If S is a protoalgebraic logic, then $\mathbf{Alg}S = \mathbf{Alg}^*S$; and if S is algebraizable, then $\mathbf{Alg}S$ is its equivalent quasivariety semantics.*

PROOF. By Proposition 2.24 we have in general that $\mathbf{Alg}^*S \subseteq \mathbf{Alg}S$. Now let $A \in \mathbf{Alg}S$ and put F_0 for its least S-filter; then $\Omega_A(F_0) = \tilde{\Omega}_A(\mathcal{F}i_S A) = Id_A$, which means that $\langle A, F_0 \rangle \in \mathbf{Matr}^*S$, that is, $A \in \mathbf{Alg}^*S$. This proves the first assertion. If moreover S is algebraizable, then by Corollary 5.3 of Blok and Pigozzi [1989a] we know that its equivalent quasivariety semantics is the class $\mathbf{Alg}^*S$; but every algebraizable sentential logic is also protoalgebraic (see Blok and Pigozzi [1989a] p. 35), and so we can apply the first part of this proof and obtain that the equivalent quasivariety semantics of S is the class $\mathbf{Alg}S$. $\dashv$

The preceding result is an important step on the way to justifying the adequacy of *considering $\mathbf{Alg}S$ as the algebraic counterpart of an arbitrary logic S*: Protoalgebraic logics are precisely those whose matrix semantics behaves reasonably well (see Blok and Pigozzi [1986], and especially Blok and Pigozzi [1992] and Czelakowski [2001a] to confirm this), and we see that in this case, the class of algebras ordinarily associated with a logic using matrix semantics, that is, $\mathbf{Alg}^*S$, coincides with our general algebraic counterpart of S. In particular, we see that if S is algebraizable, in which case its relationship with a distinguished class of algebras (its equivalent quasivariety semantics) is very strong, then this class equals

Alg$\mathcal{S}$. Proposition 3.2 also justifies our use of terms and notations originally used in the literature for restricted classes of logics, as discussed on page 36.

The converses of the two implications of Proposition 3.2 are not true in general: Take any consistent but not almost inconsistent algebraizable (thus a fortiori protoalgebraic) logic $\mathcal{S}$, and then consider its "purely inferential" version Wójcicki [1988, pp. 41 ff], here denoted as $\mathcal{S}_\emptyset$, which is defined just by $Th\mathcal{S}_\emptyset = Th\mathcal{S} \cup \{\emptyset\}$. It is straightforward to check that this defines a sentential logic which is not protoalgebraic; nevertheless **Alg**$\mathcal{S}_\emptyset$ = **Alg**$\mathcal{S}$ = **Alg**$^*\mathcal{S}$ = **Alg**$^*\mathcal{S}_\emptyset$. Another non-trivial example can be found in Section 5.4.1 on relevance logics. In Corollary 2.25 we saw that the equality **Alg**$^*\mathcal{S}$ = **Alg**$\mathcal{S}$ is also true whenever **Alg**$^*\mathcal{S}$ is a quasivariety. The results of 2.25 and 3.2 are not related, since there are protoalgebraic logics $\mathcal{S}$ such that **Alg**$^*\mathcal{S}$ is not a quasivariety (Herrmann's LJ logic in [1993b] is an example) while there are non-protoalgebraic logics $\mathcal{S}$ such that **Alg**$^*\mathcal{S}$ is a variety ($\mathcal{S}_\emptyset$, where $\mathcal{S}$ is classical logic, is but one example; a less artificial one is the logic WR described in Section 5.4.1).

We have already seen in Proposition 3.1 that the notion of protoalgebraicity can be characterized in terms of the Tarski congruence of the closure systems $(\mathcal{F}i_\mathcal{S}A)^F$ for $F \in \mathcal{F}i_\mathcal{S}A$. We will study the behaviour of the mapping $F \longmapsto (\mathcal{F}i_\mathcal{S}A)^F$ on $\mathcal{F}i_\mathcal{S}A$, and will solve specifically two questions:

– When do all full models have the form $\langle A, (\mathcal{F}i_\mathcal{S}A)^F \rangle$ for some $F \in \mathcal{F}i_\mathcal{S}A$?
– When will all the abstract logics having this form be full models?

First notice that the full models of protoalgebraic logics are determined by their theorems:

LEMMA 3.3. *Let $\mathcal{S}$ be a protoalgebraic logic. If $\mathbb{L}_1$ and $\mathbb{L}_2$ are two full models of $\mathcal{S}$ on the same algebra with $C_1(\emptyset) = C_2(\emptyset)$ then $\mathbb{L}_1 = \mathbb{L}_2$.*

PROOF. We can apply Proposition 3.1(ii) and write

$$\widetilde{\Omega}_A(\mathbb{L}_1) = \Omega_A\big(C_1(\emptyset)\big) = \Omega_A\big(C_2(\emptyset)\big) = \widetilde{\Omega}(\mathbb{L}_2)$$

and then by Theorem 2.30 it follows that $\mathbb{L}_1 = \mathbb{L}_2$.   ⊣

The following characterization of protoalgebraicity answers the first of the two questions just raised.

THEOREM 3.4. *Let $\mathcal{S}$ be any sentential logic. Then $\mathcal{S}$ is protoalgebraic if and only if all full models of $\mathcal{S}$ have the form $\langle A, (\mathcal{F}i_\mathcal{S}A)^F \rangle$ for some algebra A and some $F \in \mathcal{F}i_\mathcal{S}A$.*

PROOF. ($\Rightarrow$) Let $\mathbb{L} = \langle A, C \rangle$ be any full model of $\mathcal{S}$, and take $F = C(\emptyset)$; then obviously $C \subseteq (\mathcal{F}i_\mathcal{S}A)^F$. Since $\mathcal{S}$ is protoalgebraic, $\widetilde{\Omega}(\mathbb{L}) = \Omega_A(F)$ and thus the projection $\pi : A \rightarrow A/\Omega_A(F)$ is a bilogical morphism between $\langle A, C \rangle$

and $\langle A/\Omega_A(F), C/\Omega_A(F)\rangle$, but since $\mathbb{L}$ is a full model of $\mathcal{S}$, $C/\Omega_A(F) = \mathcal{F}i_{\mathcal{S}}(A/\Omega_A(F))$. Now take any $G \in (\mathcal{F}i_{\mathcal{S}}A)^F$. Since $F \subseteq G$ and $\mathcal{S}$ is protoalgebraic, $\Omega_A(F)$ is compatible with G, so $G = \pi^{-1}[\pi[G]]$, therefore by Proposition 1.19 $\pi[G]$ is an $\mathcal{S}$-filter on the quotient; now this implies that $\pi^{-1}[\pi[G]] \in \mathcal{C}$, that is, $G \in \mathcal{C}$. This proves that $\mathbb{L} = \langle A, (\mathcal{F}i_{\mathcal{S}}A)^F\rangle$.

($\Leftarrow$) Let $F, F' \in \mathcal{F}i_{\mathcal{S}}A$ with $F \subseteq F'$ and consider $\Omega_A(F)$: Since by Proposition 2.24 $\mathbf{Alg}^*\mathcal{S} \subseteq \mathbf{Alg}\mathcal{S}$, we know that $\Omega_A(F) \in \mathrm{Con}_{\mathbf{Alg}\mathcal{S}}A$, and by Theorem 2.30 there is some full model of $\mathcal{S}$ on A, $\mathbb{L} = \langle A, \mathcal{C}\rangle$, such that $\Omega_A(F) = \widetilde{\Omega}(\mathbb{L})$. Since $\mathbb{L}$ is a full model of $\mathcal{S}$, this implies that $\pi : A \rightarrow A/\Omega_A(F)$ is a bilogical morphism from $\mathbb{L} = \langle A, \mathcal{C}\rangle$ to $\langle A/\Omega_A(F), \mathcal{F}i_{\mathcal{S}}(A/\Omega_A(F))\rangle$; and since always $F = \pi^{-1}[\pi[F]]$, $F \in \mathcal{C}$. But by assumption there is a $G \in \mathcal{F}i_{\mathcal{S}}A$ such that $\mathcal{C} = (\mathcal{F}i_{\mathcal{S}}A)^G$; therefore, $F \supseteq G$ and as a consequence also $F' \supseteq G$, that is, $F' \in \mathcal{C}$, and this implies that $\Omega_A(F) = \widetilde{\Omega}(\mathbb{L}) = \widetilde{\Omega}_A(\mathcal{C}) \subseteq \Omega_A(F')$. We have proved that Ω_A is monotone on $\mathcal{F}i_{\mathcal{S}}A$, that is, $\mathcal{S}$ is protoalgebraic. $\dashv$

In general, for any logic $\mathcal{S}$ and any algebra A we can consider

$$\mathcal{F}i_{\mathcal{S}}^{\star}A = \left\{F \in \mathcal{F}i_{\mathcal{S}}A : \langle A, (\mathcal{F}i_{\mathcal{S}}A)^F\rangle \text{ is a full model of } \mathcal{S}\right\},$$

which is a subfamily of $\mathcal{F}i_{\mathcal{S}}A$. As a consequence of the above result we get an interesting property of protoalgebraic logics:

PROPOSITION 3.5. *If $\mathcal{S}$ is a protoalgebraic logic then for any A the Leibniz operator Ω_A is a lattice isomorphism between $\mathcal{F}i_{\mathcal{S}}^{\star}A$ and $\mathrm{Con}_{\mathbf{Alg}^*\mathcal{S}}A = \mathrm{Con}_{\mathbf{Alg}\mathcal{S}}A$.*

PROOF. The mapping $F \mapsto \langle A, (\mathcal{F}i_{\mathcal{S}}A)^F\rangle$ always maps $\mathcal{F}i_{\mathcal{S}}^{\star}A$ to $\mathcal{F}\mathcal{M}od_{\mathcal{S}}A$, is one-to-one, and satisfies that $\langle A, (\mathcal{F}i_{\mathcal{S}}A)^F\rangle \leqslant \langle A, (\mathcal{F}i_{\mathcal{S}}A)^G\rangle$ if and only if $F \subseteq G$. If moreover $\mathcal{S}$ is protoalgebraic, then Theorem 3.4 tells us that it is surjective; therefore it is an order-isomorphism between $\mathcal{F}i_{\mathcal{S}}^{\star}A$ and $\mathcal{F}\mathcal{M}od_{\mathcal{S}}A$. But by Theorem 2.30 the lattice $\mathcal{F}\mathcal{M}od_{\mathcal{S}}A$ is isomorphic, through the Tarski operator, to $\mathrm{Con}_{\mathbf{Alg}\mathcal{S}}A$, thus the composition of the two mappings is $F \mapsto \widetilde{\Omega}_A((\mathcal{F}i_{\mathcal{S}}A)^F)$ and is an order-isomorphism between $\mathcal{F}i_{\mathcal{S}}^{\star}A$ and $\mathrm{Con}_{\mathbf{Alg}\mathcal{S}}A$; using again the fact that $\mathcal{S}$ is protoalgebraic, this mapping is the same as the mapping $F \mapsto \Omega_A(F)$, that is, it is the Leibniz operator. Finally, since $\mathcal{S}$ is protoalgebraic, we can use Proposition 3.2 and conclude that $\mathbf{Alg}^*\mathcal{S} = \mathbf{Alg}\mathcal{S}$; thus $\mathrm{Con}_{\mathbf{Alg}\mathcal{S}}A = \mathrm{Con}_{\mathbf{Alg}^*\mathcal{S}}A$. $\dashv$

We will now see how the $\mathcal{S}$-filters in $\mathcal{F}i_{\mathcal{S}}^{\star}A$ can be characterized independently of the notion of full model of $\mathcal{S}$[21]. To this end, for any sentential logic $\mathcal{S}$ and any A we consider the following binary relation on $\mathcal{F}i_{\mathcal{S}}A$ (actually, the kernel of the Leibniz operator):

$$F \sim F' \iff \Omega_A(F) = \Omega_A(F').$$

Obviously Proposition 3.5 implies that when $\mathcal{S}$ is protoalgebraic at most one filter in each equivalence class belongs to $\mathcal{F}i_{\mathcal{S}}^{\star}A$; we will characterize it. Observe that when $\mathcal{S}$ is protoalgebraic each equivalence class has a minimum: If for any $F \in \mathcal{F}i_{\mathcal{S}}A$ we denote its equivalence class by $[F]$, then $\bigcap[F] \in \mathcal{F}i_{\mathcal{S}}A$ and $\Omega_A([F]) = \bigcap\{\Omega_A(G) : G \in [F]\} = \Omega_A(F)$, that is, $\bigcap[F] \in [F]$. This is the filter we look for:

PROPOSITION 3.6. *Let $\mathcal{S}$ be a protoalgebraic logic. Then for any A and any $F \in \mathcal{F}i_{\mathcal{S}}A$ the following conditions are equivalent:*

(i) *$F \in \mathcal{F}i_{\mathcal{S}}^{\star}A$, that is, $\langle A, (\mathcal{F}i_{\mathcal{S}}A)^F \rangle$ is a full model of $\mathcal{S}$;*

(ii) *F is the minimum of its equivalence class under $\sim$; and*

(iii) *$F/\Omega_A(F)$ is the least $\mathcal{S}$-filter on $A/\Omega_A(F)$.*

PROOF. (ii)⇒(iii): If $G \in \mathcal{F}i_{\mathcal{S}}(A/\Omega_A(F))$ consider $F' = \pi^{-1}[G] \cap F \in \mathcal{F}i_{\mathcal{S}}A$, where $\pi : A \to A/\Omega_A(F)$. Then $F' = \pi^{-1}[G] \cap \pi^{-1}[\pi[F]] = \pi^{-1}[G \cap \pi[F]]$, thus F' is a union of equivalence classes, that is, $\Omega_A(F)$ is compatible with F', which implies $\Omega_A(F) \subseteq \Omega_A(F')$; but on the other hand $F' \subseteq F$ and since $\mathcal{S}$ is protoalgebraic, $\Omega_A(F') \subseteq \Omega_A(F)$, so finally $\Omega_A(F) = \Omega_A(F')$. Thus $F \sim F'$ and the assumption on F implies $F \subseteq F'$, so $F = F'$. Therefore $F \subseteq \pi^{-1}[G]$ which implies $F/\Omega_A(F) = \pi[F] \subseteq G$. Therefore $F/\Omega_A(F)$ is the least $\mathcal{S}$-filter on $A/\Omega_A(F)$.

(iii)⇒(i): If $\mathcal{S}$ is protoalgebraic, we know that for any $F \in \mathcal{F}i_{\mathcal{S}}A$ the natural projection $\pi : A \to A/\Omega_A(F)$ establishes a lattice isomorphism between $(\mathcal{F}i_{\mathcal{S}}A)^F$ and $(\mathcal{F}i_{\mathcal{S}}(A/\Omega_A(F)))^{F/\Omega_A(F)}$; see page 60. Now the assumption in (iii) means that this last family is equal to $\mathcal{F}i_{\mathcal{S}}(A/\Omega_A(F))$; taking into account that $\widetilde{\Omega}_A((\mathcal{F}i_{\mathcal{S}}A)^F)$ is $\Omega_A(F)$, this means that $\langle A, (\mathcal{F}i_{\mathcal{S}}A)^F \rangle \in \mathcal{F}Mod_{\mathcal{S}}A$, that is, $F \in \mathcal{F}i_{\mathcal{S}}^{\star}A$.

(i)⇒(ii): Let $F \in \mathcal{F}i_{\mathcal{S}}^{\star}A$, and let G be the minimum of the equivalence class of F under $\sim$ (such a minimum exists because of the protoalgebraicity of $\mathcal{S}$). Using the two preceding parts of the proof we conclude that $\mathbb{L}_G = \langle A, (\mathcal{F}i_{\mathcal{S}}A)^G \rangle \in \mathcal{F}Mod_{\mathcal{S}}A$, and by assumption $\mathbb{L}_F = \langle A, (\mathcal{F}i_{\mathcal{S}}A)^F \rangle \in \mathcal{F}Mod_{\mathcal{S}}A$. But then

---

[21] These filters and their properties in protoalgebric logics have been more extensively studied in Font and Jansana [2001], where the term **Leibniz filter** was adopted, and in Jansana [2003]. See also Font, Jansana, and Pigozzi [2001] for the application of this notion in other investigations in abstract algebraic logic.

$\widetilde{\Omega}_{\mathbf{A}}(\mathbb{L}_F) = \Omega_{\mathbf{A}}(F) = \Omega_{\mathbf{A}}(G) = \widetilde{\Omega}_{\mathbf{A}}(\mathbb{L}_G)$ and by Theorem 2.30 this implies $\mathbb{L}_F = \mathbb{L}_G$, that is, $F = G$. Therefore F is the minimum of its own equivalence class under $\sim$. ⊣

One of the properties of the Leibniz operator which has an important role in some characterizations of algebraizable logics is injectiveness; in this respect the following observation may be of some interest:

PROPOSITION 3.7. *Let* S *be a protoalgebraic logic. Then* $\mathcal{F}i_S^{\star}\mathbf{A} = \mathcal{F}i_S\mathbf{A}$ *(that is, for every* $F \in \mathcal{F}i_S\mathbf{A}$, *the abstract logic* $\langle \mathbf{A}, (\mathcal{F}i_S\mathbf{A})^F \rangle$ *is a full model of* S) *if and only if the Leibniz operator* $\Omega_{\mathbf{A}}$ *is injective on* $\mathcal{F}i_S\mathbf{A}$.

PROOF. The equality $\mathcal{F}i_S^{\star}\mathbf{A} = \mathcal{F}i_S\mathbf{A}$ means that each S-filter is the only member of its own equivalence class under $\sim$, and this is equivalent to saying that $\Omega_{\mathbf{A}}(F) = \Omega_{\mathbf{A}}(G)$ implies $F = G$. ⊣

Now we can round up these results, together with some of the previous chapter, to obtain several characterizations of *the sentential logics whose full models can be completely "identified" with their filters in a natural way:*

THEOREM 3.8. *For any sentential logic* S *the following conditions are equivalent:*

(i) S *is protoalgebraic and for every* $\mathbf{A}$ *and every* $F \in \mathcal{F}i_S\mathbf{A}$, $F/\Omega_{\mathbf{A}}(F)$ *is the least* S-filter on $\mathbf{A}/\Omega_{\mathbf{A}}(F)$;

(ii) *For every* $\mathbf{A}$, *the Leibniz operator* $\Omega_{\mathbf{A}}$ *is monotone and injective on* $\mathcal{F}i_S\mathbf{A}$;

(iii) *For every* $\mathbf{A}$, *the mapping* $F \mapsto \langle \mathbf{A}, (\mathcal{F}i_S\mathbf{A})^F \rangle$ *is a bijection (and as a consequence a lattice isomorphism) between* $\mathcal{F}i_S\mathbf{A}$ *and* $\mathcal{F}Mod_S\mathbf{A}$;

(iv) *For every* $\mathbf{A}$, $\Omega_{\mathbf{A}}$ *is a lattice isomorphism between* $\mathcal{F}i_S\mathbf{A}$ *and* $\mathrm{Con}_{\mathbf{Alg}_S}\mathbf{A}$;

(v) *For every* $\mathbf{A}$, $\Omega_{\mathbf{A}}$ *is a lattice isomorphism between* $\mathcal{F}i_S\mathbf{A}$ *and* $\mathrm{Con}_{\mathbf{Alg}^*_S}\mathbf{A}$.

PROOF. (i)$\Longleftrightarrow$(ii) comes from Propositions 3.6 and 3.7.

(i)$\Rightarrow$(iii): The mapping $F \mapsto \langle \mathbf{A}, (\mathcal{F}i_S\mathbf{A})^F \rangle$ is always injective; by Proposition 3.6 the second assumption implies that for every $F \in \mathcal{F}i_S\mathbf{A}$ its image falls in $\mathcal{F}Mod_S\mathbf{A}$, and Theorem 3.4 tells us that it is surjective. Therefore it is a bijection between $\mathcal{F}i_S\mathbf{A}$ and $\mathcal{F}Mod_S\mathbf{A}$. Since by definition both this mapping and its inverse are trivially order-preserving, the mapping is a lattice isomorphism.

(iii)$\Rightarrow$(iv): In particular the mapping $F \mapsto \langle \mathbf{A}, (\mathcal{F}i_S\mathbf{A})^F \rangle$ is onto $\mathcal{F}Mod_S\mathbf{A}$, thus by Theorem 3.4 S is protoalgebraic. On the other hand, the composition of this isomorphism with that of Theorem 2.30 gives us an isomorphism from $\mathcal{F}i_S\mathbf{A}$ to $\mathrm{Con}_{\mathbf{Alg}_S}\mathbf{A}$, which now is $F \mapsto \widetilde{\Omega}(\langle \mathbf{A}, (\mathcal{F}i_S\mathbf{A})^F \rangle) = \Omega_{\mathbf{A}}(F)$ by part (iii) of Proposition 3.1, that is, it is the Leibniz operator.

(iv)$\Rightarrow$(v): We always have that $\mathrm{Con}_{\mathbf{Alg}^*_S}\mathbf{A} \subseteq \mathrm{Con}_{\mathbf{Alg}_S}\mathbf{A}$, and also that for any

$F \in \mathcal{F}i_{\mathcal{S}}A$, $\Omega_A(F) \in \mathrm{Con}_{\mathbf{Alg}^*\mathcal{S}}A$. But by the isomorphism of (iv), each element of $\mathrm{Con}_{\mathbf{Alg}\mathcal{S}}A$ is of the form $\Omega_A(F)$ for some $F \in \mathcal{F}i_{\mathcal{S}}A$, and this implies the equality $\mathrm{Con}_{\mathbf{Alg}^*\mathcal{S}}A = \mathrm{Con}_{\mathbf{Alg}\mathcal{S}}A$, and we get (v).

(v)$\Rightarrow$(ii) is trivial.   ⊣

The slight difference between items (iv) and (v) may be of some interest if one needs to use them for some logic $\mathcal{S}$ before proving that it is protoalgebraic; the reason is that until one proves this one cannot assume that the classes $\mathbf{Alg}\mathcal{S}$ and $\mathbf{Alg}^*\mathcal{S}$ are in fact the same.

The sentential logics satisfying the conditions appearing in the last Theorem deserve a name of their own:

DEFINITION 3.9. *A sentential logic $\mathcal{S}$ is **weakly algebraizable** when for any A, the Leibniz operator Ω_A is monotone and injective on $\mathcal{F}i_{\mathcal{S}}A$.*

As Theorem 3.8 shows, these logics have the outstanding property (iii) that there is a natural lattice isomorphism between their filters and their full models on a given algebra. They have been studied mainly in Czelakowski and Jansana [2000] and Czelakowski [2001a]; in addition to the behaviour of the Leibniz operator, they can be characterized by the existence of an equational logic to which they are equivalent by means of elementary definable structural translations *with parameters*. An example of a sentential logic which is weakly algebraizable but not algebraizable in the stronger sense of Blok and Pigozzi is due to Andréka and Németi, and appears in Appendix 2 of Blok and Pigozzi [1989a]. From the definition of weakly algebraizable logics, it follows that to be algebraizable they only lack the condition of continuity for Ω_A. From this fact we will obtain a new characterization of algebraizability in terms of the Tarski operator; to this end we say that, for some algebra A, the Tarski operator $\widetilde{\Omega}_A$ is **continuous** (over $\mathcal{F}Mod_{\mathcal{S}}A$) when for any upwards directed family $\{\mathbb{L}_i : i \in I\} \subseteq \mathcal{F}Mod_{\mathcal{S}}A$ we have

$$\widetilde{\Omega}_A\Big(\sup_{i \in I} \mathbb{L}_i\Big) = \bigcup_{i \in I} \widetilde{\Omega}_A(\mathbb{L}_i);$$

where directedness and the "sup" operation refer to the natural ordering between abstract logics, that is, the natural ordering between closure operators, or the inverse one between closure systems, as defined in page 18. We then have:

THEOREM 3.10. *Let $\mathcal{S}$ be a weakly algebraizable sentential logic. Then the following conditions are equivalent:*

  (i) *$\mathcal{S}$ is algebraizable;*

 (ii) *The class $\mathbf{Alg}\mathcal{S}$ is a quasivariety;*

(iii) *For any A, the Leibniz operator Ω_A is continuous on $\mathcal{F}i_S A$; and*

(iv) *For any A, the Tarski operator $\widetilde{\Omega}_A$ is continuous on $\mathcal{F}Mod_S A$.*

PROOF. It is well-known that (ii) follows from (i), and taking (ii) into account, the isomorphism established in part (iv) of Theorem 3.8 implies (i) by the characterization of algebraizability of Theorem 5.1 of Blok and Pigozzi [1989a]. The equivalence between (i) and (iii), given Theorem 3.8, is contained in Theorem 13.15 of Blok and Pigozzi [1992]. So we have only to prove the equivalence between (iii) and (iv). If for any $F \in \mathcal{F}i_S A$ we put $\Phi(F) = \langle A, (\mathcal{F}i_S A)^F \rangle$, we know that $\Omega_A = \widetilde{\Omega}_A \circ \Phi$ (because S is protoalgebraic) and thus that $\widetilde{\Omega}_A = \Omega_A \circ \Phi^{-1}$ (because Φ is a bijection, by Theorem 3.8). Now assume that Ω_A is continuous and let $\{\mathbb{L}_i : i \in I\} \subseteq \mathcal{F}Mod_S A$ be any directed family; if we put $F_i = \Phi^{-1}(\mathbb{L}_i)$ and $G = \bigcup\{F_i : i \in I\}$, then it is clear that $\{F_i : i \in I\} \subseteq \mathcal{F}i_S A$ is also a directed family and thus $G \in \mathcal{F}i_S A$; therefore $\Phi(G) = \langle A, (\mathcal{F}i_S A)^G \rangle \in \mathcal{F}Mod_S A$. Since clearly $(\mathcal{F}i_S A)^G = \bigcap\{(\mathcal{F}i_S A)^{F_i} : i \in I\}$, it easily follows that $\Phi(G) = \sup_{i \in I} \mathbb{L}_i$ and then

$$\widetilde{\Omega}_A(\sup_{i \in I} \mathbb{L}_i) = \left(\Omega_A \circ \Phi^{-1}\right)\left(\Phi(G)\right) = \Omega_A(G) = \bigcup_{i \in I} \Omega_A(F_i) = \bigcup_{i \in I} \widetilde{\Omega}_A(\mathbb{L}_i)$$

which proves that $\widetilde{\Omega}_A$ is continuous. Conversely, if we assume that $\widetilde{\Omega}_A$ is continuous and $\{F_i : i \in I\} \subseteq \mathcal{F}i_S A$ is directed, clearly the family $\{\Phi(F_i) : i \in I\}$ is also directed and

$$\Omega_A(\bigcup_{i \in I} F_i) = \widetilde{\Omega}_A(\Phi(\bigcup_{i \in I} F_i)) = \widetilde{\Omega}_A(\sup_{i \in I} \mathbb{L}_i) = \bigcup_{i \in I} \widetilde{\Omega}_A(\mathbb{L}_i) = \bigcup_{i \in I} \Omega_A(F_i)$$

which shows that Ω_A is continuous. ⊣

COROLLARY 3.11. *For any sentential logic S the following conditions are equivalent:*

(i) *S is algebraizable;*

(ii) *S is weakly algebraizable and $\mathbf{Alg}S$ is a quasivariety; and*

(iii) *For every A the mapping $F \mapsto \langle A, (\mathcal{F}i_S A)^F \rangle$ is a bijection between the sets $\mathcal{F}i_S A$ and $\mathcal{F}Mod_S A$, and the Tarski operator $\widetilde{\Omega}_A$ is continuous over $\mathcal{F}Mod_S A$.* ⊣

Therefore we see that the logics which are weakly algebraizable but not algebraizable in the sense of Blok and Pigozzi [1989a] must be such that their associated class of algebras $\mathbf{Alg}S$ is not a quasivariety. Moreover, the bijection between filters and full models of S established by the mapping $F \mapsto \langle A, (\mathcal{F}i_S A)^F \rangle$ confirms a feature of algebraizable logics that had been empirically observed earlier (and which we now know is characteristic of a larger class of logics); we will make some use of these facts later on.

Now we introduce another distinct class of sentential logics:

DEFINITION 3.12. *A sentential logic S is called **Fregean** when for any $\Gamma \in ThS$, the abstract logic S^{Γ} has the congruence property; i.e., when $\Lambda_S(\Gamma) = \widetilde{\Omega}(S^{\Gamma})$ for all $\Gamma \in ThS$*[22].

It is easy to check that every two-valued logic (i.e., every logic defined by a matrix on any two-element algebra) is Fregean. In view of the expression (1.6) of page 29, we see that S is Fregean when for any $\Gamma \in ThS$ and any $\varphi, \psi \in Fm$ it holds that

$$\text{if } \Gamma, \varphi \dashv\vdash_S \Gamma, \psi \text{ then for any } \gamma(p, \vec{q}) \in Fm, \tag{3.10}$$
$$\Gamma, \gamma(\varphi, \vec{q}) \dashv\vdash_S \Gamma, \gamma(\psi, \vec{q}).$$

So we see that these logics enjoy a very strong property of replacement of equivalents. Moreover, from (3.10) it follows that any Fregean logic satisfies the so-called **Suszko's rules** (cf. Czelakowski [1981] Theorem II.1.2 and Rautenberg [1993]): For any $\varphi, \psi, \gamma(p, \vec{q}) \in Fm$ it holds that

$$\varphi, \psi, \gamma(\varphi, \vec{q}) \vdash_S \gamma(\psi, \vec{q}).$$

From this and expression (1.1) on page 16 one can easily obtain:

PROPOSITION 3.13. *If S is a Fregean logic then the filter of each of its reduced matrices is either empty or a one-element subset.*   ⊣

The above observations suggest that attaching the name of Frege to these logics may be a reasonable choice; in Rautenberg [1981] they are called "congruential", but this term has also been used with other meanings in the literature (see for instance Blok and Pigozzi [1992]). The subclass of Fregean protoalgebraic logics has been independently introduced and studied by Pigozzi and Czelakowski (in unpublished notes[23]) in relation to the class of *Fregean varieties* of algebras

---

[22]This definition has been complemented in later literature, starting with Babyonyshev [2003] and Font [2003b], with that of the class of the *fully Fregean* logics. These are the logics S such that for every full model $\mathbb{L} = \langle A, \mathcal{C} \rangle$ of S and every $T \in \mathcal{C}$, the abstract logic $\mathbb{L}^T = \langle A, \mathcal{C}^T \rangle$ has the congruence property, that is, $\Lambda_{\mathbb{L}}(T) = \widetilde{\Omega}(\mathbb{L}^T)$. The now called *Frege hierarchy* is the classification scheme of sentential logics under the four classes defined in terms of congruence properties: the selfextensional ones, the fully selfextensional ones, the Fregean ones and the fully Fregean ones. Some results in this and the next chapters are the first steps in the clarification of the structure of the Frege hierarchy and its relations with the Leibniz hierarchy. See also Font [2006], Section 3.4.

[23]Their results have been subsequently published in Czelakowski and Pigozzi [2004a], [2004b]; see also Chapter 6 of Czelakowski [2001a].

considered in Pigozzi [1991]; such logics can be characterized in a very simple way:

PROPOSITION 3.14. *A sentential logic S is Fregean and protoalgebraic if and only if for any $\Gamma \in ThS$, $\Omega_{Fm}(\Gamma) = \Lambda_S(\Gamma)$.*

PROOF. By Definition 3.12 and Proposition 3.1, if S is both Fregean and protoalgebraic, we have that for any $\Gamma \in ThS$, $\Lambda_S(\Gamma) = \Lambda(S^{\Gamma}) = \widetilde{\Omega}(S^{\Gamma}) = \Omega_{Fm}(\Gamma)$. Conversely, if for every $\Gamma \in ThS$ we have the equalities $\Lambda(S^{\Gamma}) = \Lambda_S(\Gamma) = \Omega_{Fm}(\Gamma)$, then on the one hand Ω_{Fm} is order-preserving on ThS, that is, S is protoalgebraic, and on the other hand $\Lambda(S^{\Gamma})$ is a congruence for every $\Gamma \in ThS$, that is, S is Fregean. $\dashv$

From the definition it trivially follows that any Fregean logic is a fortiori self-extensional. That the class of Fregean logics is strictly smaller than the class of the selfextensional ones will be shown in Chapter 5 through several examples. At the end of this chapter and in Chapter 4 we will find some relationships between the class of Fregean logics and the class of the strongly selfextensional ones.

If we consider the mapping $F \mapsto \langle A, (\mathcal{F}i_S A)^F \rangle$ in the particular case where $A = Fm$, we obtain the mapping $\Gamma \mapsto S^{\Gamma}$. We will see that this mapping also has an interesting behaviour when S is Fregean and has theorems:

PROPOSITION 3.15. *If S is a Fregean logic with theorems, then the mapping $\Gamma \mapsto S^{\Gamma}$ is an order-preserving embedding of ThS into $\mathcal{F}Mod_S Fm$.*

PROOF. Observe that if $\Gamma \in ThS$ then Γ is the set of theorems of the abstract logic S^{Γ}; as a consequence, the mapping $\Gamma \mapsto S^{\Gamma}$ is one-to-one, and obviously order-preserving. It remains only to show that $S^{\Gamma} \in \mathcal{F}Mod_S Fm$, that is, putting $\theta = \widetilde{\Omega}(S^{\Gamma}) = \Lambda_S(\Gamma) = \Lambda(S^{\Gamma})$, we have to show that $(ThS^{\Gamma})/\theta = \mathcal{F}i_S(Fm/\theta)$. One half is always true, because θ is compatible with all $\Gamma' \in ThS^{\Gamma}$ and therefore $\Gamma'/\theta \in \mathcal{F}i_S(Fm/\theta)$. Now let F be any S-filter on Fm/θ; then $\pi^{-1}[F]$ is also an S-filter on Fm, that is, $\pi^{-1}[F] \in ThS$, and we have only to show that it contains Γ: Since we are assuming that S has theorems, we can always take any $\varphi \in \Gamma$ and any $\psi \in \Gamma \cap \pi^{-1}[F]$. Then $\langle \varphi, \psi \rangle \in \Lambda_S(\Gamma) = \theta$, so $\pi(\varphi) = \pi(\psi) \in F$ which implies $\varphi \in \pi^{-1}[F]$; that is, $\Gamma \subseteq \pi^{-1}[F]$. This shows that $\pi^{-1}[F] \in ThS^{\Gamma}$, therefore $F \in (ThS^{\Gamma})/\theta$ as was to be proved. $\dashv$

The assumption that S has theorems cannot be dropped from this result. The reason is the fact that if S does not have theorems, then no full model of S can have them; as a consequence, for any non-empty theory Γ, the abstract logic S^{Γ} cannot be a full model of S. At this point one could conjecture that the mapping $\Gamma \mapsto (S^{\Gamma})_{\emptyset}$ (using the notation introduced in page 62) would solve this problem, but we have found a proof only in a very restricted case: A sentential logic, or more generally an abstract logic, is called ***pseudo-axiomatic*** (Łoś and Suszko [1958]) when it has no theorems but has a smallest non-empty theory. Then:

PROPOSITION 3.16. *If S is a pseudo-axiomatic Fregean logic, then the mapping $\Gamma \mapsto (S^{\Gamma})_{\emptyset}$ is an order-preserving embedding of ThS into $\mathcal{F}\mathcal{M}od_S \boldsymbol{Fm}$.*

PROOF. Very similar to that of Proposition 3.15. Observe that if $\Gamma \in ThS$ then $(S^{\Gamma})_{\emptyset}$ is also pseudo-axiomatic and Γ is its smallest non-empty theory. The mapping is obviously one-to-one and order-preserving. We have to show that $(S^{\Gamma})_{\emptyset} \in \mathcal{F}\mathcal{M}od_S \boldsymbol{Fm}$. If $\Gamma = \emptyset$ this is trivially true since then $(S^{\Gamma})_{\emptyset} = S$, so let us suppose that Γ is non-empty. Observe that $\widetilde{\Omega}((S^{\Gamma})_{\emptyset}) = \widetilde{\Omega}(S^{\Gamma})$; thus we can take $\theta = \widetilde{\Omega}(S^{\Gamma}) = \Lambda_S(\Gamma) = \Lambda(S^{\Gamma})$, and show that $Th((S^{\Gamma})_{\emptyset})/\theta = \mathcal{F}i_S(\boldsymbol{Fm}/\theta)$. One half is always true, because θ is compatible with all non-empty $\Gamma' \in Th((S^{\Gamma})_{\emptyset})$ and therefore $\Gamma'/\theta \in \mathcal{F}i_S(\boldsymbol{Fm}/\theta)$; while by assumption the empty set is in $\mathcal{F}i_S(\boldsymbol{Fm}/\theta)$. The converse is proved with the same construction as in the proof of 3.15, because for a non-empty $F \in \mathcal{F}i_S(\boldsymbol{Fm}/\theta)$, the set $\Gamma \cap \pi^{-1}[F]$ is also non-empty, because it contains the least non-empty theory of S, and everything works similarly. The case $F = \emptyset$ is trivial. $\dashv$

However, pseudo-axiomatic logics are rather unnatural, and so this result is of not much help. There are Fregean logics without theorems satisfying the conclusion of Proposition 3.16, but at present it seems that an ad-hoc proof using particular characterizations of their full models is needed in every case; see for instance in Section 5.1.1 the case of $CPC_{\wedge\vee}$, the $\{\wedge, \vee\}$-fragment of CPC.

If moreover the sentential logic S is protoalgebraic, then we can say more about the mapping initially considered:

PROPOSITION 3.17. *If S is a Fregean protoalgebraic logic with theorems, then the mapping $\Gamma \mapsto S^{\Gamma}$ is an isomorphism between the lattices ThS and $\mathcal{F}\mathcal{M}od_S \boldsymbol{Fm}$.*

PROOF. In view of Proposition 3.15, we need only to show that the mapping $\Gamma \mapsto S^{\Gamma}$ is onto $\mathcal{F}\mathcal{M}od_S \boldsymbol{Fm}$. But this is a consequence of the assumption that S is protoalgebraic, by Theorem 3.4 applied to the case $\boldsymbol{A} = \boldsymbol{Fm}$. $\dashv$

In this case, the assumption that S has theorems can be substituted by the assumption that S is not the almost inconsistent logic, since it is known that the latter is the only protoalgebraic logic without theorems. And this is also an exception to the conclusion: If $ThS = \{\emptyset, Fm\}$ then the mapping $\Gamma \mapsto S^{\Gamma}$ is not into $\mathcal{F}\mathcal{M}od_S \boldsymbol{Fm}$, since S^{Fm}, which is the inconsistent logic, does not belong to

$\mathcal{FMod}_S\, Fm$, because it has theorems while S does not. Actually, the full models of the almost inconsistent logic are all abstract logics $\langle A, C \rangle$ with $C = \{\emptyset, A\}$.

Note that the isomorphism proved in Proposition 3.17 is a particular case of the one obtained in part (iii) of Theorem 3.8 under different assumptions. As a consequence we find an alternative proof of the following result contained in Czelakowski [1992][24]. A sentential logic is *regularly algebraizable* if it is algebraizable and the filter of any of its reduced matrices is a one-element subset. These logics have also been studied in Herrmann [1993b], [1993a] under the name of *1-equivalential logics*.

THEOREM 3.18 (Czelakowski, Pigozzi). *Every Fregean protoalgebraic logic with theorems is regularly algebraizable.*

PROOF. By Proposition 3.1, $\Omega_{Fm}(\Gamma) = \widetilde{\Omega}(S^{\Gamma})$ for every $\Gamma \in ThS$; therefore the composition of the isomorphisms of Theorems 3.17 and 2.30 results to be the mapping Ω_{Fm}, which becomes an isomorphism from ThS to $\mathrm{Con}_{\mathsf{Alg}S}\, Fm$. By Proposition 3.14, $\Omega_{Fm} = \Lambda_S$, the Frege operator, which by Proposition 2.38 always preserves unions of directed families of theories. So Ω_{Fm}, on ThS, is injective, order-preserving, and preserves unions of directed families. This is exactly the "first intrinsic characterization" of algebraizability found in Theorem 4.2 of Blok and Pigozzi [1989a]; therefore we conclude that S is algebraizable. Now let $\langle A, F \rangle$ be a reduced matrix for S. Since S has theorems, F is non-empty, and then Proposition 3.13 tells us that F is a singleton. Therefore S is regularly algebraizable.   ⊣

This result shows the strength of being Fregean: these logics must be regularly algebraizable, or else they cannot be even protoalgebraic, leaving the almost inconsistent logic aside. So in particular we see that the only Fregean logic which is equivalential or finitely equivalential without being algebraizable is the almost inconsistent one. This confirms one of the claims made in Font [1993] concerning the classification of sentential logics outlined there.

As an application of this theorem the relationship between strongly selfextensional and Fregean sentential logics is partly clarified in the following results.

PROPOSITION 3.19. *Every Fregean protoalgebraic logic is strongly selfextensional.*

PROOF. If S does not have theorems, then it is the almost inconsistent logic; as we observed before, its full models are $\langle A, C \rangle$ with $C = \{\emptyset, A\}$, and hence they have the congruence property, that is, the logic S is strongly selfextensional. Now

---

[24] This has been subsequently published as Theorem 6.2.2 of Czelakowski [2001a], and as Theorem 2.18 of Czelakowski and Pigozzi [2004a].

let us assume that $\mathcal{S}$ has theorems. Then we can use the result of Corollary 5.5 of Czelakowski [1992], which, expressed in our notation, says that, under the same assumptions, for any A and any $F \in \mathcal{F}i_{\mathcal{S}}A$, $\Omega_A(F) = \Lambda\big((\mathcal{F}i_{\mathcal{S}}A)^F\big)$. This implies that the abstract logic $\langle A, (\mathcal{F}i_{\mathcal{S}}A)^F \rangle$ has the congruence property. But from Theorem 3.4 it follows that all the full models of $\mathcal{S}$ have this form, for some $F \in \mathcal{F}i_{\mathcal{S}}A$. Therefore, all the full models of $\mathcal{S}$ have the congruence property, that is, $\mathcal{S}$ is strongly selfextensional.   ⊣

PROPOSITION 3.20. *Let $\mathcal{S}$ be a strongly selfextensional sentential logic. Then the following conditions are equivalent:*

(i) *$\mathcal{S}$ is Fregean, protoalgebraic, and has theorems.*

(ii) *$\mathcal{S}$ is algebraizable.*

(iii) *$\mathcal{S}$ is weakly algebraizable.*

PROOF. Part (i)⇒(ii) is contained in Theorem 3.18, and part (ii)⇒(iii) is trivial, so let us prove (iii)⇒(i): If $\mathcal{S}$ is weakly algebraizable, then Ω_{Fm} is monotone and injective on $Th\mathcal{S}$, thus in particular $\mathcal{S}$ is protoalgebraic. If we take $A = Fm$ in Theorem 3.8, we find that every axiomatic extension of $\mathcal{S}$ is a full model of $\mathcal{S}$. Hence if $\mathcal{S}$ is strongly selfextensional, these axiomatic extensions have the congruence property, that is, $\mathcal{S}$ is Fregean. Finally, since $\Omega_{Fm}(\emptyset) = \Omega_{Fm}(Fm) = Fm \times Fm$ and Ω_{Fm} is injective on $Th\mathcal{S}$, we have that $\emptyset \notin Th\mathcal{S}$, therefore $\mathcal{S}$ has theorems.   ⊣

From the preceding results we highlight two things: First, that among strongly selfextensional logics, being weakly algebraizable implies being algebraizable in the stronger sense of Blok and Pigozzi [1989a]. Second, by combining Proposition 3.19 and Proposition 3.20, we find:

COROLLARY 3.21. *Let $\mathcal{S}$ be any weakly algebraizable sentential logic. Then $\mathcal{S}$ is strongly selfextensional if and only if $\mathcal{S}$ is Fregean.*   ⊣

The coincidence of strongly selfextensional and Fregean logics holds a fortiori inside the class of algebraizable logics. On the other hand, the assumption of weak algebraizability cannot be dropped from 3.21: in Sections 5.3, 5.4.4 and 5.4.3 we present some examples of protoalgebraic logics that are strongly selfextensional but not Fregean; and in Sections 5.1.2, 5.1.3 and 5.4.1 several non-protoalgebraic logics being strongly selfextensional but not Fregean are presented.

Concerning these classifications we can highlight:

OPEN PROBLEM. *Is there a logic that is Fregean but not strongly selfextensional?*[25]

---

[25]Such an example is presented in Babyonyshev [2003].

Note that by Proposition 3.19 a logic of this kind should be non-protoalgebraic. As a consequence of Theorem 4.28 in the next chapter, such a logic cannot have a conjunction, either.

CHAPTER 4

ABSTRACT LOGICS AS MODELS
OF GENTZEN SYSTEMS

In this chapter we will introduce the issue of considering abstract logics as models of Gentzen systems, and characterize a kind of sentential logics whose full models can be described as, essentially, the models of some Gentzen system. We will also relate our study with the theory of the *algebraization of Gentzen systems*; this generalization of Blok and Pigozzi's theory of the algebraization of sentential logics was begun in Rebagliato and Verdú [1993] for some particular cases, and the general theory has started to be developed in Rebagliato and Verdú [1995][26]. We will treat some general material in Section 4.1, and in Sections 4.2 and 4.3 two particular cases will be studied, where things behave quite well. As a by-product we will get interesting results about properties of sentential logics; in particular, the open problem presented in Chapter 2 (page 48) will be solved for two important classes of logics.

Note that while in the literature Gentzen systems are mostly used to reason about their derivable sequents, in principle nothing prevents us from considering the relation of derivability of a sequent from other sequents; it is in this sense that we consider Gentzen systems in this monograph, that is, as a kind of *sequential logic*, a relation of consequence operating on sequents rather than on formulas, whose axioms are called *initial sequents* and whose theorems are called *derivable sequents* in the standard terminology. As a matter of fact, many particular *Gentzen calculi* exist in the literature having some particular axioms (i.e., initial sequents) besides the sequent $\varphi \vdash \varphi$, so one can just generalize this procedure. We will use the symbol $\vdash_{\mathfrak{G}}$ to denote this relation of derivability; thus when we write

$$\{\Gamma_i \vdash \varphi_i : i \in I\} \vdash_{\mathfrak{G}} \Gamma \vdash \varphi$$

we mean that there is a derivation of the sequent $\Gamma \vdash \varphi$ using the rules of the

---

[26]Later papers that have somehow continued the same trend are Pynko [1999] and Raftery [2006].

Gentzen system $\mathfrak{G}$ whose initial sequents are among the initial sequents (or axioms) of $\mathfrak{G}$ or in the set $\{\Gamma_i \vdash \varphi_i : i \in I\}$; this is more classically (and more graphically) expressed by saying that the rule

$$\frac{\{\Gamma_i \vdash \varphi_i : i \in I\}}{\Gamma \vdash \varphi}$$

is a *derived rule* of $\mathfrak{G}$. Although the tree-like notation may be more intuitive to talk about sequents, we will use the alternative notation with $\mathord{\sim}_{\mathfrak{G}}$ more often, partly to save space, and partly because we are not dealing with proof-theoretic issues that might require the classical notation.

Since our goal is to treat Gentzen systems only in order to study the sentential logics that they define and such that their models (in a certain, natural, sense) are abstract logics (to be able to compare them with the full models of the sentential logic), we will not deal with completely arbitrary Gentzen systems, but with those satisfying the so-called *structural rules*. Moreover, the *sequents* we will treat will have a *finite set* of formulas, rather than a sequence or a multiset, on the left-hand side of the turnstile (the symbol $\vdash$) and just *one* formula on its right-hand side; as a consequence, there is no point in considering the rules of Exchange and Contraction. The reader should thus bear in mind that what we call a *Gentzen system* in this chapter is a restricted case of what this term commonly describes in the literature.

4.1. Gentzen systems and their models

For our needs, we will take a *sequent* of formulas to be a pair $\langle \Gamma, \varphi \rangle$ where Γ is a finite (possibly empty) set of formulas and φ is a formula; tradition compels us to write $\Gamma \vdash \varphi$ instead of $\langle \Gamma, \varphi \rangle$, and to use the customary notational abbreviations like $\Gamma, \psi \vdash \varphi$ for $\Gamma \cup \{\psi\} \vdash \varphi$, etc. We will consider the set $\mathrm{Seq}(\boldsymbol{Fm})$ of all sequents, and the set $\mathrm{Seq}^\circ(\boldsymbol{Fm}) = \{\Gamma \vdash \varphi \in \mathrm{Seq}(\boldsymbol{Fm}) : \Gamma \neq \emptyset\}$ of all sequents with non-empty left-hand side. We will use boldface Greek letters to stand for sequents (lowercase: $\boldsymbol{\delta}, \boldsymbol{\sigma}$) and sets of sequents (uppercase: $\boldsymbol{\Delta}, \boldsymbol{\Sigma}$).

DEFINITION 4.1. *A **Gentzen system of type** ω (resp. **of type** ω°) is a pair $\mathfrak{G} = \langle \boldsymbol{Fm}, \mathord{\sim}_{\mathfrak{G}} \rangle$ where $\mathord{\sim}_{\mathfrak{G}}$ is a finitary and structural consequence relation on the set $\mathrm{Seq}(\boldsymbol{Fm})$ (resp. on the set $\mathrm{Seq}^\circ(\boldsymbol{Fm})$) which in addition satisfies the following **structural rules**:*

(Axiom) $\emptyset \mathrel{\vdash\!\sim}_{\mathfrak{G}} \varphi \vdash \varphi$ *for every* $\varphi \in Fm$.

(Weakening) $\Gamma \vdash \varphi \mathrel{\vdash\!\sim}_{\mathfrak{G}} \Gamma, \psi \vdash \varphi$ *for every* $\Gamma \cup \{\varphi, \psi\} \subseteq Fm$.

(Cut) $\quad \{\Gamma \vdash \varphi\,,\,\Gamma, \varphi \vdash \psi\} \mathrel{\big|\!\sim}_{\mathfrak{G}} \Gamma \vdash \psi$ *for every* $\Gamma \cup \{\varphi, \psi\} \subseteq Fm$.

In this definition, by a finitary consequence relation on the sets $\mathrm{Seq}(Fm)$ or $\mathrm{Seq}^{\circ}(Fm)$ we understand the obvious generalization to sequents of the notion of finitary consequence relation of a sentential logic: $\mathrel{\big|\!\sim}_{\mathfrak{G}}$ is a binary relation between sets of sequents and sequents satisfying conditions (S1) to (S4) of page 25 with formulas replaced by sequents; and for it to be structural means the generalization of condition (S5) by extending homomorphisms to sequents in the obvious way: If $\{\Gamma_i \vdash \varphi_i : i \in I\} \mathrel{\big|\!\sim}_{\mathfrak{G}} \Gamma \vdash \varphi$ then for any homomorphism h of Fm into itself, $\{h[\Gamma_i] \vdash h(\varphi_i) : i \in I\} \mathrel{\big|\!\sim}_{\mathfrak{G}} h[\Gamma] \vdash h(\varphi)$. Several other notions are similarly extended from the formula concept to a sequent concept. If $\emptyset \mathrel{\big|\!\sim}_{\mathfrak{G}} \Gamma \vdash \varphi$ then we say that the sequent $\Gamma \vdash \varphi$ is a *derivable sequent* of $\mathfrak{G}$.

Note that since by definition all our Gentzen systems have Weakening, they also have as a derived rule a more general form of the Cut rule, which written in tree-like form is

$$\frac{\Gamma \vdash \varphi \qquad \Delta, \varphi \vdash \psi}{\Gamma, \Delta \vdash \psi}\,.$$

We will often refer to applications of this rule by the same term "Cut rule".

If Σ and Δ are sets of sequents, then $\Sigma \mathrel{\big|\!\sim}_{\mathfrak{G}} \Delta$ means that $\Sigma \mathrel{\big|\!\sim}_{\mathfrak{G}} \delta$ does hold for every $\delta \in \Delta$, and $\Sigma \mathrel{\dashv\!\big|\!\sim}_{\mathfrak{G}} \Delta$ means that both $\Sigma \mathrel{\big|\!\sim}_{\mathfrak{G}} \Delta$ and $\Delta \mathrel{\big|\!\sim}_{\mathfrak{G}} \Sigma$ hold.

For any Gentzen system $\mathfrak{G}$ we denote by $\mathrm{Seq}(\mathfrak{G})$ either $\mathrm{Seq}(Fm)$ if $\mathfrak{G}$ is of type ω or $\mathrm{Seq}^{\circ}(Fm)$ if $\mathfrak{G}$ is of type ω°, and we call *sequents of* $\mathfrak{G}$ the elements of $\mathrm{Seq}(\mathfrak{G})$. This consideration of Gentzen systems of different types[27] is a simplification of the terminology introduced in Rebagliato and Verdú [1993]; our sequents of type ω are called "of type $(\omega, \{1\})$" in Rebagliato and Verdú [1993], [1995], while those of type ω° are called "of type $(\omega \smallsetminus \{0\}, \{1\})$". The consideration of two different kinds of Gentzen systems is motivated by the need to treat Gentzen systems for all kinds of sentential logics, with or without theorems, in a uniform way; this may become clearer in the comments after the following definition.

DEFINITION 4.2. *Let $\mathfrak{G}$ be a Gentzen system. **The sentential logic defined by** $\mathfrak{G}$ is the sentential logic $\langle Fm, \vdash_{\mathfrak{G}} \rangle$ where the consequence relation $\vdash_{\mathfrak{G}}$ is defined in the following way: For all $\Gamma \subseteq Fm$, $\varphi \in Fm$,*

$$\Gamma \vdash_{\mathfrak{G}} \varphi \iff \text{there is a finite } \Delta \subseteq \Gamma \text{ such that } \emptyset \mathrel{\big|\!\sim}_{\mathfrak{G}} \Delta \vdash \varphi.$$

---

[27] The closely related notion of *trace* has been introduced in Raftery [2006] to allow for a greater generalization of these ideas.

*If S is a sentential logic, then we say that $\mathfrak{G}$ is **adequate for** S when S is the sentential logic defined by $\mathfrak{G}$ (that is, $\vdash_{\mathfrak{G}} = \vdash_S$) and moreover either S has theorems and $\mathfrak{G}$ is of type ω, or S does not have theorems and $\mathfrak{G}$ is of type ω°.*

Note that the first part of this definition really gives a sentential logic because we are assuming that $\mathfrak{G}$ satisfies the structural rules (see Definition 4.1). We can summarize the second part of Definition 4.2 by saying that a Gentzen system $\mathfrak{G}$ is adequate for a sentential logic S when S is the sentential logic defined by $\mathfrak{G}$, and $\mathfrak{G}$ is of the specified type according to whether S has or has not theorems. The following observations are straightforward:

1. If $\mathfrak{G}$ is of type ω° then $\vdash_{\mathfrak{G}}$ has no theorems.
2. If $\mathfrak{G}$ is of type ω then its restriction $\mathfrak{G}^{\circ}$ to $\mathrm{Seq}^{\circ}(\boldsymbol{Fm})$ is also a Gentzen system, and it is of type ω°.
3. If $\mathfrak{G}$ is of type ω and $\vdash_{\mathfrak{G}}$ has no theorems, then $\vdash_{\mathfrak{G}} = \vdash_{\mathfrak{G}^{\circ}}$.

These facts tell us that for our purposes there is no point in using sequents of the form $\emptyset \vdash \varphi$ when the sentential logic defined by the Gentzen system has no theorems. This is the motivation behind our use of Gentzen systems of two different types in the notion of *adequacy* of a Gentzen system for a sentential logic depending on whether the logic has or has not theorems; see Definition 4.2.

For any sentential logic S there is a general way of obtaining a Gentzen system $\mathfrak{G}$ that is trivially adequate for S: Take it as being of type ω or ω° according to whether S has or does not have theorems, take the structural rule (Axiom) of Definition 4.1 and the elements of the set $\{\Gamma \vdash \varphi \in \mathrm{Seq}(\mathfrak{G}) : \Gamma \vdash_S \varphi\}$ as axioms, and (Weakening) and (Cut) as the only rules. It is straightforward to check that $\vdash_{\mathfrak{G}} = \vdash_S$. However, with this definition we cannot guarantee that $\mathfrak{G}$ has any of the metalogical properties of S as a derivable rule; for instance in Font and Verdú [1991], pp. 403–404, it is shown that the Gentzen system so obtained from $\mathrm{CPC}_{\wedge\vee}$, the $\{\wedge, \vee\}$-fragment of classical logic, does not have the Property of Disjunction (see also Section 5.1.1). We can thus say that the notion of adequacy just defined is too weak for our purposes. A better link will be established on the basis of the following notion.

DEFINITION 4.3. *An abstract logic $\mathbb{L} = \langle \boldsymbol{A}, C \rangle$ is a **model of a Gentzen system** $\mathfrak{G}$ when for any family of sequents $\{\Gamma_i \vdash \varphi_i : i \in I\} \cup \{\Gamma \vdash \varphi\} \subseteq \mathrm{Seq}(\mathfrak{G})$ such that $\{\Gamma_i \vdash \varphi_i : i \in I\} \vdash_{\mathfrak{G}} \Gamma \vdash \varphi$ it holds that for any $h \in \mathrm{Hom}(\boldsymbol{Fm}, \boldsymbol{A})$ such that $h(\varphi_i) \in C\big(h[\Gamma_i]\big)$ for all $i \in I$, also $h(\varphi) \in C\big(h[\Gamma]\big)$.*

This notion was introduced in Font and Verdú [1991], Definition 2.11, for finitary abstract logics, with the closure operator replaced by its associated consequence relation. This notion of model parallels the notion of matrix model of a sentential logic; thus it is natural to expect that the models on the formula algebra correspond to the "theories" of the Gentzen system. Let us call a set $\Sigma \subseteq \mathrm{Seq}(\mathfrak{G})$ a *closed set of* $\mathfrak{G}$ when Σ is closed under the relation $\vdash_{\mathfrak{G}}$. If for any such set and any $\Gamma \subseteq Fm$ we define the set

$$C_\Sigma(\Gamma) = \{\varphi \in Fm : \text{ there is a finite } \Delta \subseteq \Gamma \text{ such that } \Delta \vdash \varphi \in \Sigma\}$$

then it is easy to see that C_Σ is a finitary closure operator on Fm that is a finitary model of $\mathfrak{G}$. Conversely, given any $\langle Fm, C\rangle$ model of $\mathfrak{G}$ on Fm, the set

$$\Sigma_C = \{\Gamma \vdash \varphi \in \mathrm{Seq}(\mathfrak{G}) : \varphi \in C(\Gamma)\}$$

is a closed set of $\mathfrak{G}$. It is straightforward to check the following facts:

PROPOSITION 4.4.

(1) $\langle Fm, C\rangle$ *is a finitary model of* $\mathfrak{G}$ *on* Fm *iff* Σ_C *is a closed set of* $\mathfrak{G}$ *and* $C = C_{\Sigma_C}$.

(2) Σ *is a closed set of* $\mathfrak{G}$ *iff* $\langle Fm, C_\Sigma\rangle$ *is a finitary model of* $\mathfrak{G}$ *and* $\Sigma = \Sigma_{C_\Sigma}$.

(3) *The abstract logic* $\langle Fm, \vdash_{\mathfrak{G}}\rangle$ *is the smallest model of* $\mathfrak{G}$ *on* Fm *and it coincides with* $\langle Fm, C_\Sigma\rangle$ *where* Σ *is the set of derivable sequents of* $\mathfrak{G}$.

(4) *(Completeness)* $\{\Gamma_i \vdash \varphi_i : i \in I\} \vdash_{\mathfrak{G}} \Gamma \vdash \varphi$ *if and only if for every model* $\mathbb{L} = \langle A, C\rangle$ *of* $\mathfrak{G}$ *and every* $h \in \mathrm{Hom}(Fm, A)$, *if* $h(\varphi_i) \in C(h[\Gamma_i])$ *for all* $i \in I$ *then* $h(\varphi) \in C(h[\Gamma])$. $\dashv$

PROPOSITION 4.5. *Let* $\mathfrak{G}$ *be a Gentzen system and let* $\mathbb{L}, \mathbb{L}'$ *be two abstract logics such that there is a bilogical morphism between them. Then* $\mathbb{L}$ *is a model of* $\mathfrak{G}$ *if and only if* $\mathbb{L}'$ *is a model of* $\mathfrak{G}$. *In particular, an abstract logic* $\mathbb{L}$ *is a model of* $\mathfrak{G}$ *if and only if its reduction* $\mathbb{L}^*$ *is a model of* $\mathfrak{G}$.

PROOF. Assume that h is a bilogical morphism from $\mathbb{L}$ onto $\mathbb{L}'$. We prove that $\mathbb{L}$ is a model of $\mathfrak{G}$ if and only if $\mathbb{L}'$ is a model of $\mathfrak{G}$.

($\Rightarrow$) Suppose that $\mathbb{L}$ is a model of $\mathfrak{G}$, assume that $\{\Gamma_i \vdash \varphi_i : i \in I\} \vdash_{\mathfrak{G}} \Gamma \vdash \varphi$ and let $g \in \mathrm{Hom}(Fm, A')$ be such that $g(\varphi_i) \in C'(g[\Gamma_i])$ for all $i \in I$. Since h is onto, there is $f \in \mathrm{Hom}(Fm, A)$ satisfying $h \circ f = g$. Thus we have $h(f(\varphi_i)) \in C'(h[f[\Gamma_i]])$ for each $i \in I$, and therefore, using that h is a bilogical morphism, $f(\varphi_i) \in h^{-1}[C'(h[f[\Gamma_i]])] = C(f[\Gamma_i])$. Hence, since $\mathbb{L}$ is a model of $\mathfrak{G}$, this implies $f(\varphi) \in C(f[\Gamma])$ which implies $h(f(\varphi)) \in h[C(f[\Gamma])] = C'(h[f[\Gamma]])$, that is, $g(\varphi) \in C'(g[\Gamma])$. Thus also $\mathbb{L}'$ is a model of $\mathfrak{G}$.

($\Leftarrow$) Suppose that $\mathbb{L}'$ is a model of $\mathfrak{G}$, assume that $\{\Gamma_i \vdash \varphi_i : i \in I\} \vdash_{\mathfrak{G}} \Gamma \vdash \varphi$ and let $g \in \mathrm{Hom}(Fm, A)$ be such that $g(\varphi_i) \in C(g[\Gamma_i]) = h^{-1}[C'(h[g[\Gamma_i]])]$ for all $i \in I$. This implies $h(g(\varphi_i)) \in C'(h[g[\Gamma_i]])$ for all $i \in I$, and so also $h(g(\varphi)) \in C'(h[g[\Gamma]])$; therefore $g(\varphi) \in h^{-1}[C'(h[g[\Gamma]])] = C(g[\Gamma])$, which proves that $\mathbb{L}$ is a model of $\mathfrak{G}$. $\dashv$

DEFINITION 4.6. *For any abstract logic* $\mathbb{L} = \langle A, C \rangle$, *the **finitary part of** $\mathbb{L}$ is the abstract logic* $\mathbb{L}_{\text{fin}} = \langle A, C_{\text{fin}} \rangle$, *where* C_{fin} *is the strongest finitary closure operator weaker than* C.

Recall that C_{fin} always exists and is defined by the expression

$$C_{\text{fin}}(X) = \bigcup \{ C(Y) : Y \subseteq X \, , \, Y \text{ finite} \},$$

see for instance Wójcicki [1988] Section 1.2.2. Thus $\mathbb{L}$ is finitary if and only if $\mathbb{L} = \mathbb{L}_{\text{fin}}$. We present here some properties of the construction $\mathbb{L} \mapsto \mathbb{L}_{\text{fin}}$ that will be needed in the sequel:

PROPOSITION 4.7. *If* $\mathbb{L}$ *is an abstract logic, then* $\widetilde{\Omega}(\mathbb{L}) = \widetilde{\Omega}(\mathbb{L}_{\text{fin}})$ *and* $\Lambda(\mathbb{L}) = \Lambda(\mathbb{L}_{\text{fin}})$. *As a consequence* $\mathbb{L}$ *is reduced if and only if* $\mathbb{L}_{\text{fin}}$ *is reduced, and* $\mathbb{L}$ *has the congruence property if and only if* $\mathbb{L}_{\text{fin}}$ *has it. Moreover, if* $\mathfrak{G}$ *is any Gentzen system, then* $\mathbb{L}$ *is a model of* $\mathfrak{G}$ *if and only if* $\mathbb{L}_{\text{fin}}$ *is.*

PROOF. From the expression we have just given for defining C_{fin} it follows that for any finite $X \subseteq A$, $C(X) = C_{\text{fin}}(X)$; in particular for any $a \in A$, $C(a) = C_{\text{fin}}(a)$. This immediately implies $\Lambda(\mathbb{L}) = \Lambda(\mathbb{L}_{\text{fin}})$, and by the characterization (1.3) of $\widetilde{\Omega}(\mathbb{L})$ on page 19 it also implies $\widetilde{\Omega}(\mathbb{L}) = \widetilde{\Omega}(\mathbb{L}_{\text{fin}})$. From these equalities the two stated consequences follow trivially. Finally, since being a model of a Gentzen system $\mathfrak{G}$ involves only finite sets of formulas, the first observation implies that $\mathbb{L}$ will be a model of $\mathfrak{G}$ if and only if $\mathbb{L}_{\text{fin}}$ is. $\dashv$

Having defined a very general notion of model of a Gentzen system, it is natural to single out the algebraic reducts of the reduced models as a class of algebras naturally associated with the Gentzen system:

DEFINITION 4.8. *Let* $\mathfrak{G}$ *be a Gentzen system and* A *be an algebra. We say that* A *is a $\mathfrak{G}$-**algebra** when* A *is the algebraic reduct of a reduced model of* $\mathfrak{G}$. *The class of all $\mathfrak{G}$-algebras will be denoted by* **Alg**$\mathfrak{G}$.

Notice that by Proposition 4.7 we can assume without loss of generality that the models considered in this definition are finitary.

LEMMA 4.9. *Let* $\mathfrak{G}$ *be a Gentzen system adequate for* S. *Then every model of* $\mathfrak{G}$ *is a model of* S, *and* **Alg**$\mathfrak{G} \subseteq$ **Alg**S.

PROOF. Suppose that $\Gamma \vdash_S \varphi$. By assumption there is a finite $\Delta \subseteq \Gamma$ such that $\emptyset \hspace{0.1em}\vdash_{\mathfrak{G}} \Delta \vdash \varphi$. Since the left part of this relation is vacuously satisfied by every model $\mathbb{L} = \langle A, C \rangle$ of $\mathfrak{G}$ and any $h \in \text{Hom}(\boldsymbol{Fm}, A)$, we have $h(\varphi) \in C(h[\Delta]) \subseteq C(h[\Gamma])$. That is, $\mathbb{L}$ is a model of S. Therefore every model of $\mathfrak{G}$ is a model of S, and then every reduced model of $\mathfrak{G}$ is a reduced model of S. By taking algebraic reducts and using Proposition 2.19 we obtain $\mathbf{Alg}\mathfrak{G} \subseteq \mathbf{Alg}S$. ⊣

Now, suppose that a sentential logic S has an adequate Gentzen system $\mathfrak{G}$, and consider the two following ways for associating a class of algebras and a class of abstract logics with S: The standard one of S-algebras and full models of S, and the new one of $\mathfrak{G}$-algebras and the (finitary) models of $\mathfrak{G}$. In principle this second method may depend on the $\mathfrak{G}$ chosen; for instance if $\mathfrak{G}$ is the "trivial" one described just before Definition 4.3 the models of $\mathfrak{G}$ are all the models of S, not the full models. One of the main tasks of this chapter is to find conditions for the existence of a Gentzen system $\mathfrak{G}$ such that both methods give the same result. In order to investigate this issue we introduce the idea of a Gentzen system whose finitary models are precisely the full models of the sentential logic; with some technical adjustments, this gives rise to the following definition:

DEFINITION 4.10. *Let $\mathfrak{G}$ be a Gentzen system and S be a sentential logic. We say that $\mathfrak{G}$ is **strongly adequate**[28] for S when one of the two following conditions holds:*

(A) *S has theorems, $\mathfrak{G}$ is of type ω and for every abstract logic $\mathbb{L}$ of the similarity type of $\boldsymbol{Fm}$, $\mathbb{L}$ is a full model of S iff $\mathbb{L}$ is a finitary model of $\mathfrak{G}$.*

(B) *S does not have theorems, $\mathfrak{G}$ is of type ω°, and for every abstract logic $\mathbb{L}$ of the similarity type of $\boldsymbol{Fm}$, $\mathbb{L}$ is a full model of S iff $\mathbb{L}$ is a finitary model of $\mathfrak{G}$ without theorems.*

PROPOSITION 4.11. *If $\mathfrak{G}$ is a Gentzen system strongly adequate for a sentential logic S then $\mathfrak{G}$ is adequate for S.*

PROOF. We only have to prove that $\vdash_S = \vdash_\mathfrak{G}$. If $\Gamma \vdash_\mathfrak{G} \varphi$, there is some finite $\Gamma_0 \subseteq \Gamma$ such that $\emptyset \mathrel{\vdash\!\!\!\sim}_\mathfrak{G} \Gamma_0 \vdash \varphi$; from this it follows that $\Gamma_0 \vdash_S \varphi$ because S itself is a full model of S, so by assumption it is a model of $\mathfrak{G}$, and thus also $\Gamma \vdash_S \varphi$. Therefore $\vdash_\mathfrak{G} \subseteq \vdash_S$. We also know that $\langle \boldsymbol{Fm}, \vdash_\mathfrak{G} \rangle$ is a model of $\mathfrak{G}$, it is finitary, and it does not have theorems if S has none either; therefore, by assumption, it is a full model of S. Since by Proposition 2.10 S is the weakest full model of S on $\boldsymbol{Fm}$, it follows that $\vdash_S \subseteq \vdash_\mathfrak{G}$, thus completing the proof. ⊣

The notion of strong adequacy has been defined in terms of the two classes of abstract logics, associated with S and $\mathfrak{G}$ respectively. The following characterization, in terms of the two classes of algebras associated with them, will be especially useful:

---

[28]This notion has been further and more deeply investigated in Font, Jansana, and Pigozzi [2001], [2006], where the alternative and slightly more descriptive term *fully adequate* has been adopted.

PROPOSITION 4.12. *Let $\mathfrak{G}$ be a Gentzen system and S be a sentential logic. Then $\mathfrak{G}$ is strongly adequate for S if and only if the following conditions hold:*

(1) $\mathbf{Alg}S = \mathbf{Alg}\mathfrak{G}$;

(2) *For every $A \in \mathbf{Alg}S$, the abstract logic $\langle A, \mathcal{F}i_S A \rangle$ is the only finitary and reduced model of $\mathfrak{G}$ (having no theorems, if S hasn't) on A; and*

(3) *Either S has theorems and $\mathfrak{G}$ is of type ω, or S has no theorems and $\mathfrak{G}$ is of type ω°.*

PROOF. ($\Rightarrow$) If $\mathfrak{G}$ is strongly adequate for S then (3) holds by definition. Moreover by Lemma 4.9 $\mathbf{Alg}\mathfrak{G} \subseteq \mathbf{Alg}S$. If $A \in \mathbf{Alg}S$ then we know that $\langle A, \mathcal{F}i_S A \rangle$ is a full model of S, and that it is reduced; the assumption implies that it is a reduced model of $\mathfrak{G}$, therefore $A \in \mathbf{Alg}\mathfrak{G}$, thus completing the proof of (1). Finally, by assumption, finitary and reduced models of $\mathfrak{G}$ (having no theorems, if S has none) are exactly the reduced full models of S; if $A \in \mathbf{Alg}S$ then $\langle A, \mathcal{F}i_S A \rangle$ is such a reduced full model, and by the Isomorphism Theorem 2.30 it is the only full model of S on A to be reduced. This proves (2).

($\Leftarrow$) Let $\mathbb{L} = \langle A, C \rangle$ be any abstract logic. Then $\mathbb{L}$ is a full model of S iff $A^* \in \mathbf{Alg}S$ and $C^* = \mathcal{F}i_S A^*$. But by (1) and (2) this is equivalent to saying that $A^* \in \mathbf{Alg}\mathfrak{G}$ and $\langle A^*, C^* \rangle$ is a reduced finitary model of $\mathfrak{G}$ (without theorems if S has none), and this by Proposition 4.5 is equivalent to saying that $\langle A, C \rangle$ is a finitary model of $\mathfrak{G}$ (without theorems if S has none). Taking (3) into account, we conclude that $\mathfrak{G}$ is strongly adequate for S. $\dashv$

It is natural to ask whether *every sentential logic has a strongly adequate Gentzen system*. The general answer is negative; a counterexample is given in Section 5.3.1. On the other hand, *if there is a Gentzen system $\mathfrak{G}$ strongly adequate for a sentential logic S, then it is unique*; this is so because S is characterized by its full models (Theorem 2.22) while a Gentzen system is also characterized by its models (Proposition 4.4). Note that we are talking of the uniqueness of Gentzen systems as consequence relations on sequents, and not as specific presentations of the system. Obviously the same consequence $\vdash_\mathfrak{G}$ can have different presentations in terms of axioms and rules, which might have different properties from the proof-theoretical point of view (and maybe some authors would prefer to speak of them as different *calculi*).

While a sentential logic may have several adequate Gentzen systems defining it (see Section 5.2.1 for an example), we will see in the next two sections that under reasonable hypotheses a strongly adequate Gentzen system exists, and is therefore unique; it is a distinguished object naturally associated with the sentential logic. Our results will be based on another kind of relationship between a Gentzen system and a class of algebras. It is the relation considered in *the theory of algebraization of Gentzen systems* developed in Rebagliato and Verdú [1993], [1995], which closely parallels the theory of algebraization of sentential logics due to Blok and Pigozzi [1989a], [1992], [200x]. The main tool in these theories is the following relation of consequence between equations.

DEFINITION 4.13. *For each class of algebras* **K**, *the relation of* ***equational consequence relative to* K** *is the relation* $\models_{\mathbf{K}} \subseteq P\big(\mathrm{Eq}(\boldsymbol{Fm})\big) \times \mathrm{Eq}(\boldsymbol{Fm})$ *defined as:*

$$\{\varphi_i \approx \psi_i : i \in I\} \models_{\mathbf{K}} \varphi \approx \psi \iff \textit{For every } \boldsymbol{A} \in \mathbf{K} \textit{ and every } \vec{a} \textit{ in } \boldsymbol{A},$$

$$\textit{if } \boldsymbol{A} \models \varphi_i \approx \psi_i \, [\vec{a}] \; \forall i \in I, \textit{ then } \boldsymbol{A} \models \varphi \approx \psi \, [\vec{a}].$$

Following usual conventions, we write $\boldsymbol{A} \models \varphi \approx \psi \, [\vec{a}]$ to mean that $\varphi^{\boldsymbol{A}}(\vec{a}) = \psi^{\boldsymbol{A}}(\vec{a})$, that is, $h(\varphi) = h(\psi)$ for any homomorphism $h \in \mathrm{Hom}(\boldsymbol{Fm}, \boldsymbol{A})$ that maps the relevant variables to the sequence $\vec{a}$. If $E \subseteq \mathrm{Eq}(\boldsymbol{Fm})$ then $\{\varphi_i \approx \psi_i : i \in I\} \models_{\mathbf{K}} E$ means that for every $\varphi \approx \psi \in E$, $\{\varphi_i \approx \psi_i : i \in I\} \models_{\mathbf{K}} \varphi \approx \psi$; the symbol $\dashv\models_{\mathbf{K}}$ also has the obvious meaning.

If **K** is the class of all algebras of the given type, then $\models_{\mathbf{K}}$ is in fact the restriction to equations of the ordinary consequence of first-order logic in a language having the algebraic operations of our similarity type as functional symbols, and equality as the only relational symbol. The consequence $\models_{\mathbf{K}}$, whose closed sets are called "the equational theories of **K**" in Blok and Pigozzi [1989a], should not be confused with the ordinary "equational logic"; actually the theories of $\models_{\mathbf{K}}$ which are closed under substitution are the equational theories, in the ordinary sense, associated with subvarieties of the variety generated by **K**. Note that $\models_{\mathbf{K}}$ always satisfies the following rules:

(*Symmetry*) $\varphi \approx \psi \models_{\mathbf{K}} \psi \approx \varphi.$

(*Transitivity*) $\{\varphi \approx \psi, \psi \approx \eta\} \models_{\mathbf{K}} \varphi \approx \eta.$

(*Congruence*) $\{\varphi_i \approx \psi_i : i < n\} \models_{\mathbf{K}} \varpi\varphi_0 \ldots \varphi_{n-1} \approx \varpi\psi_0 \ldots \psi_{n-1}$
                for every basic operation ϖ, where n is the arity of ϖ.

The rule we have called *Congruence* is equivalent to the *Replacement* rule; these, plus the Rule of Substitution, are the rules of Birkhoff calculus. If **K** is a *quasivariety* then $\models_{\mathbf{K}}$ can be axiomatized by taking all equations valid in **K** as axioms, and the following rules of inference: the three rules mentioned above plus one rule of the form $\{\varphi_i \approx \psi_i : i < n\} \models_{\mathbf{K}} \varphi \approx \psi$ for each quasi-equation of the form $\varphi_0 \approx \psi_0 \,\&\, \ldots \,\&\, \varphi_{n-1} \approx \psi_{n-1} \Rightarrow \varphi \approx \psi$ that is valid in **K**; if this class if a *variety* then the latter rules are not necessary.

In the following definition we use the notation $P_\omega^\circ(A)$ to denote the set of all finite and non-empty subsets of an arbitrary set A.

DEFINITION 4.14. *Let $\mathfrak{G}$ be a Gentzen system. A **translation from sequents into equations** is any mapping* $\mathbf{t} : \mathrm{Seq}(\mathfrak{G}) \to P_\omega^\circ(\mathrm{Eq}(\boldsymbol{Fm}))$*; this mapping is extended to arbitrary sets of sequents by defining* $\mathbf{t}(\Sigma) = \bigcup\{\mathbf{t}(\sigma) : \sigma \in \Sigma\}$*. Similarly, a **translation from equations into sequents** is any mapping* $\mathbf{s} : \mathrm{Eq}(\boldsymbol{Fm}) \to P_\omega^\circ(\mathrm{Seq}(\mathfrak{G}))$*, and if* E *is a set of equations then we define* $\mathbf{s}(E) = \bigcup\{\mathbf{s}(\varphi \approx \psi) : \varphi \approx \psi \in E\}$*.*

If **K** *is a class of algebras and* $\mathbf{t}$ *and* $\mathbf{s}$ *are translations as above, then* $\mathfrak{G}$ *is* $(\mathbf{t}, \mathbf{s})$-***equivalent to the equational consequence*** $\models_{\mathbf{K}}$ *when the following two conditions are satisfied:*

(Eq1) $\{\Gamma_i \vdash \varphi_i : i \in I\} \mathrel{\vdash\!\!\!\sim_{\mathfrak{G}}} \Gamma \vdash \varphi \iff \mathbf{t}\bigl(\{\Gamma_i \vdash \varphi_i : i \in I\}\bigr) \models_{\mathbf{K}} \mathbf{t}(\Gamma \vdash \varphi)$

(Eq2) $\varphi \approx \psi =\!\!\models_{\mathbf{K}} \mathbf{t}\bigl(\mathbf{s}(\varphi \approx \psi)\bigr)$

The lack of "symmetry" in the definition is easily resolved; there is in fact a complete symmetry regarding the behaviour of both translations:

PROPOSITION 4.15. *A Gentzen system* $\mathfrak{G}$ *is* $(\mathbf{t}, \mathbf{s})$-*equivalent to* $\models_{\mathbf{K}}$ *if and only if the following two conditions are satisfied:*

(Eq3) $\{\varphi_i \approx \psi_i : i \in I\} \models_{\mathbf{K}} \varphi \approx \psi \iff \mathbf{s}\bigl(\{\varphi_i \approx \psi_i : i \in I\}\bigr) \mathrel{\vdash\!\!\!\sim_{\mathfrak{G}}} \mathbf{s}(\varphi \approx \psi)$

(Eq4) $\Gamma \vdash \varphi \mathrel{\dashv\!\!\!\vdash\!\!\!\sim_{\mathfrak{G}}} \mathbf{s}\bigl(\mathbf{t}(\Gamma \vdash \varphi)\bigr)$

PROOF. To prove (Eq3) just apply $\mathbf{t}$ to both sides of its right-hand part, and then use first (Eq1) and after use (Eq2). And (Eq4) is true iff $\mathbf{t}(\Gamma \vdash \varphi) =\!\!\models_{\mathbf{K}} \mathbf{t}\bigl(\mathbf{s}(\mathbf{t}(\Gamma \vdash \varphi))\bigr)$, by (Eq1), and this is true just because of (Eq2). In a similar way one proves that (Eq3) and (Eq4) together imply both (Eq1) and (Eq2).   ⊣

In Rebagliato and Verdú [1993], [1995] a class of algebras **K** is called the ***equivalent algebraic semantics of a Gentzen system*** $\mathfrak{G}$ (which is then called ***algebraizable***) when $\mathfrak{G}$ is, in our terminology, $(\mathbf{t}, \mathbf{s})$-equivalent to $\models_{\mathbf{K}}$ and the two translations are, roughly speaking, *finite* and *structural*; this means that each translation is definable by substitutions from a finite set of equations and sequents, respectively, which are the translations of basic sequents and equations (those made only of variables). This extension of Blok and Pigozzi's concept of algebraizability can be applied to Gentzen systems that are adequate for logics which are not algebraizable in the sense of Blok and Pigozzi [1989a]. In this chapter we are going to use these notions only in the case where the translation from equations into sequents has the following precise form:

DEFINITION 4.16. *The translation* $\mathbf{sq} : \mathrm{Eq}(\boldsymbol{Fm}) \to P_\omega^\circ(\mathrm{Seq}(\mathfrak{G}))$ *is the mapping defined by*

$$\mathbf{sq}(\varphi \approx \psi) = \{\varphi \vdash \psi, \; \psi \vdash \varphi\}.$$

The use of this translation in equivalences between Gentzen systems and equational consequences is intimately connected with the congruence property. Let us state formally what this property means when applied to a Gentzen system:

DEFINITION 4.17. *We say that a Gentzen system $\mathfrak{G}$* **satisfies the congruence rules** *when for each basic operation ϖ of the similarity type it holds that*

$$\{\varphi_i \vdash \psi_i, \, \psi_i \vdash \varphi_i : i < n\} \mathrel{\vdash_{\mathfrak{G}}} \varpi\varphi_0 \ldots \varphi_{n-1} \vdash \varpi\psi_0 \ldots \psi_{n-1},$$

where n is the arity of the operation.

Trivially, if a Gentzen system satisfies the congruence rules then all its models have the congruence property. In the next two results we see some of the connections just mentioned, which we will use later on.

PROPOSITION 4.18. *If a Gentzen system $\mathfrak{G}$ is $(\mathbf{t}, \mathbf{sq})$-equivalent to $\models_{\mathbf{K}}$ for some class $\mathbf{K}$ of algebras and some translation $\mathbf{t}$, then $\mathfrak{G}$ satisfies the congruence rules. If moreover $\mathfrak{G}$ is adequate for some sentential logic $\mathcal{S}$, then $\mathcal{S}$ is selfextensional and the variety generated by the class $\mathbf{K}$ is the variety $\mathbf{K}_{\mathcal{S}}$ generated by the Lindenbaum-Tarski algebra of $\mathcal{S}$.*

PROOF. If we apply the translation $\mathbf{sq}$ to the congruence rules for $\models_{\mathbf{K}}$, we obtain exactly the congruence rules for $\mathfrak{G}$ as stated in Definition 4.17. Now assume that $\mathfrak{G}$ is adequate for some sentential logic $\mathcal{S}$, and that $\varphi_i \dashv\vdash_{\mathcal{S}} \psi_i$ for $i < n$; this means that $\emptyset \mathrel{\vdash_{\mathfrak{G}}} \{\varphi_i \vdash \psi_i, \, \psi_i \vdash \varphi_i : i < n\}$, and from this, using the congruence rules for $\mathfrak{G}$ and Cut, it follows that $\emptyset \mathrel{\vdash_{\mathfrak{G}}} \varpi\varphi_0 \ldots \varphi_{n-1} \vdash \varpi\psi_0 \ldots \psi_{n-1}$ and $\emptyset \mathrel{\vdash_{\mathfrak{G}}} \varpi\psi_0 \ldots \psi_{n-1} \vdash \varpi\varphi_0 \ldots \varphi_{n-1}$, and therefore that $\varpi\varphi_0 \ldots \varphi_{n-1} \dashv\vdash_{\mathcal{S}} \varpi\psi_0 \ldots \psi_{n-1}$. Thus $\mathcal{S}$ has the congruence property, that is, it is selfextensional. Finally, an equation $\varphi \approx \psi$ holds in $\mathbf{K}$ iff $\emptyset \models_{\mathbf{K}} \varphi \approx \psi$, but by (Eq3) for $\mathbf{sq}$ this is equivalent to $\emptyset \mathrel{\vdash_{\mathfrak{G}}} \{\varphi \vdash \psi, \psi \vdash \varphi\}$, which is equivalent to $\varphi \dashv\vdash_{\mathcal{S}} \psi$ because $\mathfrak{G}$ is adequate for $\mathcal{S}$; but $\mathcal{S}$ is selfextensional, hence Proposition 2.43 tells us that this is equivalent to saying that the equation $\varphi \approx \psi$ holds in the variety $\mathbf{K}_{\mathcal{S}}$. $\dashv$

A partial converse to the preceding result, which will be useful in the next sections, is the following: under some conditions one half of (Eq3), necessary for proving the equivalence between a Gentzen system and an equational consequence, holds:

PROPOSITION 4.19. *Assume that a Gentzen system $\mathfrak{G}$ satisfies the congruence rules and is adequate for a sentential logic $\mathcal{S}$. If $\{\varphi_i \approx \psi_i : i \in I\} \models_{\mathbf{K}_{\mathcal{S}}} \varphi \approx \psi$ then $\mathbf{sq}(\{\varphi_i \approx \psi_i : i \in I\}) \mathrel{\vdash_{\mathfrak{G}}} \mathbf{sq}(\varphi \approx \psi).$*

PROOF. The same argument of Proposition 4.18 proves that if $\mathfrak{G}$ satisfies the congruence rules and is adequate for S then S is selfextensional. Therefore, by 2.43, if $\varphi \approx \psi$ is an equation valid in $\mathbf{K}_S$ then $\varphi \dashv\vdash_S \psi$, and so $\emptyset \vdash_{\mathfrak{G}} \mathbf{sq}(\varphi \approx \psi)$, because $\mathfrak{G}$ is adequate for S. Moreover observe that $\mathfrak{G}$ satisfies the $\mathbf{sq}$-translations of the rules of $\models_{\mathbf{K}_S}$: Symmetry because actually $\mathbf{sq}(\varphi \approx \psi) = \mathbf{sq}(\psi \approx \varphi)$; Transitivity because of Cut, and Congruence (or Replacement) because by assumption $\mathfrak{G}$ satisfies the congruence rules. Therefore by an easy inductive argument, from a proof in $\models_{\mathbf{K}_S}$ of $\varphi \approx \psi$ from equations in $\{\varphi_i \approx \psi_i : i \in I\}$ we obtain a proof in $\mathfrak{G}$ of $\mathbf{sq}(\varphi \approx \psi)$ from sequents in $\mathbf{sq}(\{\varphi_i \approx \psi_i : i \in I\})$. $\dashv$

4.2. Selfextensional logics with Conjunction

The main goals of this section are to prove that for logics with Conjunction (i.e., that satisfy the Property of Conjunction, PC, introduced in Section 2.4) the notion of strong selfextensionality reduces to the much simpler one of selfextensionality, that any logic having these properties has a strongly adequate Gentzen system $\mathfrak{G}$ equivalent to $\models_{\mathbf{Alg}\mathfrak{G}}$ by two specific translations $\mathbf{t}_\wedge$ and $\mathbf{sq}$, and that the associated class of algebras is always a variety. These properties tell us that selfextensional logics with Conjunction are very well behaved; this adds to the extensive study of Fregean protoalgebraic logics with Conjunction in Section 6.5 of Czelakowski [2001a] and in Czelakowski and Pigozzi [2004a][29].

We begin by proving a sufficient condition for a logic with the PC to have a strongly adequate Gentzen system.

PROPOSITION 4.20. *Let S be a sentential logic with the PC, and let $\mathfrak{G}$ be a Gentzen system such that the following conditions are satisfied:*

(1) *$\mathfrak{G}$ is adequate for S.*
(2) *$\mathfrak{G}$ is $(\mathbf{t}, \mathbf{sq})$-equivalent to $\models_{\mathbf{Alg}\mathfrak{G}}$ for some translation $\mathbf{t}$.*
(3) *$\mathbf{Alg}\mathfrak{G}$ is a variety.*

Then $\mathfrak{G}$ is strongly adequate for S.

PROOF. We will show that the three conditions of Proposition 4.12 are satisfied. Condition 4.12(3) holds because $\mathfrak{G}$ is adequate for S. For the same reason, and by Lemma 4.9, $\mathbf{Alg}\mathfrak{G} \subseteq \mathbf{Alg}S$. Moreover, assumptions (1) and (2) allow us to apply Proposition 4.18 for $\mathbf{K} = \mathbf{Alg}\mathfrak{G}$ and conclude that $\mathbf{K}_S$, which contains $\mathbf{Alg}S$ by Proposition 2.26, is the variety generated by $\mathbf{Alg}\mathfrak{G}$. But by assumption

---

[29]Further investigations on selfextensional logics with Conjunction are contained in Jansana [2006], where some of the results in this section are obtained by essentially different methods.

(3) this variety is **Alg℘** itself, hence **Alg$S$** $\subseteq$ **Alg℘**, and therefore **Alg℘** = **Alg$S$**, which is condition 4.12(1). To show condition 4.12(2) let $A \in$ **Alg℘** and let $\mathbb{L} = \langle A, C \rangle$ be any finitary reduced model of $℘$ (having no theorems if S has none). Since by Proposition 4.18 $℘$ satisfies the congruence rules, $\mathbb{L}$ has the congruence property, and by 4.9 is a model of S, so by Proposition 2.46 it is a full model of S; but since it is reduced we obtain $C = \mathcal{F}i_S A$, which completes the proof of 4.12(2). Therefore, $℘$ is strongly adequate for S. ⊣

The interest of this sufficient condition is that it rests almost completely on properties of the Gentzen system, and the only relationship between it and the sentential logic that has to be proved is that the Gentzen system is adequate for it; therefore it can be especially useful to obtain strongly adequate Gentzen systems of logics for whose filters or full models a nice, direct characterization has not been found; actually, a characterization of the full models follows from strong adequacy, by definition. We will make use of this Proposition in several of the examples analyzed in Section 5.1, and also in the proof of the main result of this section. To this end we will show that there are specific $℘$ and t satisfying the assumptions of Proposition 4.20, provided that S is selfextensional and has the PC.

First we introduce the translation:

DEFINITION 4.21. *Let S be any sentential logic with the PC. The translation* $t_\wedge$ *from* $\mathrm{Seq}^\circ(\boldsymbol{Fm})$ *to* $\mathrm{Eq}(\boldsymbol{Fm})$ *is defined as follows:*

$$t_\wedge(\Sigma \vdash \varphi) = \big\{ (\textstyle\bigwedge\Sigma) \wedge \varphi \approx \textstyle\bigwedge\Sigma \big\},$$

where $\bigwedge\Sigma$ *stands for* $\big(((\varphi_{i_1} \wedge \varphi_{i_2}) \wedge \ldots) \wedge \varphi_{i_n} \big)$ *if* $\Sigma = \{\varphi_{i_1}, \ldots, \varphi_{i_n}\}$ *with* $i_1 < i_2 < \cdots < i_n$ *and* $n \geqslant 2$, *taking for granted a fixed enumeration of the set of all formulas* $Fm = \{\varphi_i : i \in \omega\}$, *while* $\bigwedge\{\varphi_i\} = \varphi_i$.

If moreover S has theorems then the translation can be extended to the whole set of sequents $\mathrm{Seq}(\boldsymbol{Fm})$ *by selecting a fixed theorem τ of S and defining*

$$t_\wedge(\emptyset \vdash \varphi) = \{\varphi \approx \tau\}.$$

Actually, since S has the PC, it will not matter which enumeration and which position of the parentheses in the expression $\bigwedge\Sigma$ is chosen for the above definition. Also note that in fact this translation can be defined independently of S if we choose τ as a fixed formula (but in the applications it will be a theorem of S).

As noted in Section 2.4, the fact that a logic S has the PC can be expressed by saying that the three following sequents

$$\{\varphi, \psi\} \vdash \varphi \wedge \psi, \quad \varphi \wedge \psi \vdash \varphi \quad \text{and} \quad \varphi \wedge \psi \vdash \psi \qquad (4.11)$$

are Hilbert-style rules of S. Therefore, if S has the PC then these three sequents must be derivable sequents of any Gentzen system $\mathfrak{G}$ adequate for S, and as a consequence every model of this $\mathfrak{G}$ will have the PC. Moreover, using Cut, one can easily prove that a Gentzen system $\mathfrak{G}$ has the sequents in (4.11) as derivable sequents if and only if the usual rules for introduction of Conjunction to both sides of the turnstile

$$\frac{\Gamma \vdash \varphi}{\bigwedge \Gamma \vdash \varphi} \qquad \text{and} \qquad \frac{\Gamma \vdash \varphi \quad \Gamma \vdash \psi}{\Gamma \vdash \varphi \wedge \psi} \qquad (4.12)$$

are derivable rules of $\mathfrak{G}$. Bearing all this in mind we prove a very general result, which can be seen as another partial converse to the first part of Proposition 4.18, for logics with the PC; moreover, it will be used when we show that the assumptions in Proposition 4.20 are satisfied.

PROPOSITION 4.22. *Let S be a sentential logic with the PC and let $\mathfrak{G}$ be a Gentzen system adequate for S and satisfying the congruence rules. Then $\mathfrak{G}$ is $(\mathbf{t}_\wedge, \mathbf{sq})$-equivalent to $\models_{\mathsf{Alg}\mathfrak{G}}$.*

PROOF. We begin by proving that $\varphi \approx \psi \dashv\models_{\mathsf{Alg}\mathfrak{G}} \mathbf{t}_\wedge(\mathbf{sq}(\varphi \approx \psi))$, which is condition (Eq2) of Definition 4.14; in our case, this means that we have to prove that $\varphi \approx \psi \dashv\models_{\mathsf{Alg}\mathfrak{G}} \{\varphi \wedge \psi \approx \varphi, \psi \wedge \varphi \approx \psi\}$. For any $A \in \mathsf{Alg}\mathfrak{G}$ there is some closure operator C over A such that the abstract logic $\mathbb{L} = \langle A, C \rangle$ is a reduced model of $\mathfrak{G}$. This abstract logic will have the PC as well, and the congruence property by the assumption that $\mathfrak{G}$ satisfies the congruence rules; this implies that $C(a) = C(b)$ holds if and only if $a = b$. Since $C(a \wedge b) = C(a, b) = C(b \wedge a)$, we obtain $a \wedge b = b \wedge a$ for all $a, b \in A$, and this implies that the equation $\varphi \wedge \psi \approx \psi \wedge \varphi$ holds in $\mathsf{Alg}\mathfrak{G}$, therefore $\{\varphi \wedge \psi \approx \varphi, \psi \wedge \varphi \approx \psi\} \models_{\mathsf{Alg}\mathfrak{G}} \varphi \approx \psi$. If $a = b$ then $C(a \wedge b) = C(a, b) = C(a)$ and $C(b \wedge a) = C(b, a) = C(b)$ thus $a \wedge b = a$ and $b \wedge a = b$; this shows that $\varphi \approx \psi \models_{\mathsf{Alg}\mathfrak{G}} \{\varphi \wedge \psi \approx \varphi, \psi \wedge \varphi \approx \psi\}$. Therefore condition (Eq2) is proved.

To prove condition (Eq1) we must prove that

$$\{\Gamma_i \vdash \varphi_i : i \in I\} \mathrel{\vdash\!\sim}_\mathfrak{G} \Gamma \vdash \varphi \Leftrightarrow \mathbf{t}_\wedge(\{\Gamma_i \vdash \varphi_i : i \in I\}) \models_{\mathsf{Alg}\mathfrak{G}} \mathbf{t}_\wedge(\Gamma \vdash \varphi).$$

($\Rightarrow$) Let $A \in \mathsf{Alg}\mathfrak{G}$, and take any C over A such that the abstract logic $\mathbb{L} = \langle A, C \rangle$ is a reduced model of $\mathfrak{G}$; this abstract logic will have the PC and the congruence property as well. Let $\vec{a}$ be a sequence of elements of A such that for each $i \in I$, $A \models \mathbf{t}_\wedge(\Gamma_i \vdash \varphi_i) [\vec{a}]$. If $\Gamma_i \neq \emptyset$, this means $A \models (\bigwedge \Gamma_i) \wedge \varphi_i \approx \bigwedge \Gamma_i [\vec{a}]$, therefore $C\big(\big((\bigwedge \Gamma_i) \wedge \varphi_i\big)^{\boldsymbol{A}}(\vec{a})\big) = C(\bigwedge \Gamma_i^{\boldsymbol{A}}(\vec{a}))$ and by the PC $\varphi_i^{\boldsymbol{A}}(\vec{a}) \in C(\Gamma_i^{\boldsymbol{A}}(\vec{a}))$. If $\Gamma_i = \emptyset$ then we have $A \models \tau \approx \varphi_i [\vec{a}]$, so $C(\tau^{\boldsymbol{A}}(\vec{a})) = C(\varphi_i^{\boldsymbol{A}}(\vec{a}))$; but since $\emptyset \vdash_S \tau$ and $\mathfrak{G}$ is adequate for S, we have $\emptyset \mathrel{\vdash\!\sim}_\mathfrak{G} \emptyset \vdash \tau$, and since $\mathbb{L} = \langle A, C \rangle$ is a model of $\mathfrak{G}$, we conclude that $\varphi_i^{\boldsymbol{A}}(\vec{a}) \in C(\tau^{\boldsymbol{A}}(\vec{a})) =$

$C(\emptyset) = C\big(\Gamma_i^{\boldsymbol{A}}(\vec{a})\big)$. We see that in every case $\varphi_i^{\boldsymbol{A}}(\vec{a}) \in C\big(\Gamma_i^{\boldsymbol{A}}(\vec{a})\big)$ for all $i \in I$. Since $\mathbb{L} = \langle \boldsymbol{A}, C \rangle$ is a model of $\mathfrak{G}$, we obtain $\varphi^{\boldsymbol{A}}(\vec{a}) \in C\big(\Gamma^{\boldsymbol{A}}(\vec{a})\big)$. Now if $\Gamma = \emptyset$ this implies that $C\big(\varphi^{\boldsymbol{A}}(\vec{a})\big) = C\big(\tau^{\boldsymbol{A}}(\vec{a})\big)$, which gives $\boldsymbol{A} \models \tau \approx \varphi \, [\vec{a}]$. If, on the other hand, $\Gamma \neq \emptyset$, then we get $C\big(\big((\bigwedge \Gamma) \wedge \varphi\big)^{\boldsymbol{A}}(\vec{a})\big) = C\big((\bigwedge \Gamma)^{\boldsymbol{A}}(\vec{a})\big)$ which implies that $\boldsymbol{A} \models (\bigwedge \Gamma) \wedge \varphi \approx \bigwedge \Gamma \, [\vec{a}]$. So in both cases we have proved that $\boldsymbol{A} \models \boldsymbol{t}_{\wedge}(\Gamma \vdash \varphi) \, [\vec{a}]$.

($\Leftarrow$): Let $\boldsymbol{\Sigma}$ be the closed set of $\mathrel{\vdash}_{\mathfrak{G}}$ generated by the set $\{\Gamma_i \vdash \varphi_i : i \in I\}$. By Proposition 4.4 the abstract logic $\mathbb{L}_{\Sigma} = \langle \boldsymbol{Fm}, C_{\Sigma} \rangle$ is a model of $\mathfrak{G}$, so by assumption $\mathbb{L}_{\Sigma}$ has the PC and the congruence property. As a consequence, $\widetilde{\boldsymbol{\Omega}}(\mathbb{L}_{\Sigma}) = \boldsymbol{\Lambda}(\mathbb{L}_{\Sigma}) = \big\{ \langle \varphi, \psi \rangle : C_{\Sigma}(\varphi) = C_{\Sigma}(\psi) \big\}$. Now suppose that $\Gamma_i \neq \emptyset$. Since by construction and the PC we have that $\varphi_i \in C_{\Sigma}(\bigwedge \Gamma_i)$, it follows that $C_{\Sigma}\big((\bigwedge \Gamma_i) \wedge \varphi_i\big) = C_{\Sigma}(\bigwedge \Gamma_i)$, that is, $\langle (\bigwedge \Gamma_i) \wedge \varphi_i, \bigwedge \Gamma_i \rangle \in \widetilde{\boldsymbol{\Omega}}(\mathbb{L}_{\Sigma})$; this implies that $\boldsymbol{Fm}/\widetilde{\boldsymbol{\Omega}}(\mathbb{L}_{\Sigma}) \models (\bigwedge \Gamma_i) \wedge \varphi_i \approx \bigwedge \Gamma_i \, [\pi]$ where π is the interpretation defined by the natural projection onto the quotient. If, on the other hand, $\Gamma_i = \emptyset$ then $C_{\Sigma}(\varphi_i) = C_{\Sigma}(\emptyset)$; this tells us that $\mathcal{S}$ must have theorems, so if τ is the theorem selected for the translation, $\emptyset \vdash \tau \in \boldsymbol{\Sigma}$, which implies $\tau \in C_{\Sigma}(\emptyset)$, and thus $C_{\Sigma}(\varphi_i) = C_{\Sigma}(\tau)$. This implies that $\boldsymbol{Fm}/\widetilde{\boldsymbol{\Omega}}(\mathbb{L}_{\Sigma}) \models \tau \approx \varphi_i \, [\pi]$, as before. Thus for all $i \in I$ we have that $\boldsymbol{Fm}/\widetilde{\boldsymbol{\Omega}}(\mathbb{L}_{\Sigma}) \models \boldsymbol{t}_{\wedge}(\Gamma_i \vdash \varphi_i) \, [\pi]$. Since $\boldsymbol{Fm}/\widetilde{\boldsymbol{\Omega}}(\mathbb{L}_{\Sigma}) \in \mathbf{Alg}\mathfrak{G}$, the assumption implies that $\boldsymbol{Fm}/\widetilde{\boldsymbol{\Omega}}(\mathbb{L}_{\Sigma}) \models \boldsymbol{t}_{\wedge}(\Gamma \vdash \varphi) \, [\pi]$. Now a similar process in the opposite direction, distinguishing the cases Γ empty and Γ non-empty, leads to the proof that $\varphi \in C_{\Sigma}(\Gamma)$. Therefore we have proved that $\{\Gamma_i \vdash \varphi_i : i \in I\} \mathrel{\vdash}_{\mathfrak{G}} \Gamma \vdash \varphi$. $\dashv$

In the following definition we associate a Gentzen system with every selfextensional logic; however, we will use it only for the ones with the PC, for which we will prove that it is the Gentzen system we are looking for.

DEFINITION 4.23. *Let $\mathcal{S}$ be a selfextensional logic. Then the Gentzen system $\mathfrak{G}_{\mathcal{S}}$ is defined by the following axioms and rules on $\mathrm{Seq}(\mathfrak{G}_{\mathcal{S}})$, which is $\mathrm{Seq}(\boldsymbol{Fm})$ or $\mathrm{Seq}^{\circ}(\boldsymbol{Fm})$ depending on whether $\mathcal{S}$ has or does not have theorems:*

(1) *The "proper axioms" $\Gamma \vdash \varphi$, for all $\Gamma \vdash \varphi \in \mathrm{Seq}(\mathfrak{G}_{\mathcal{S}})$ such that $\Gamma \vdash_{\mathcal{S}} \varphi$.*
(2) *The "structural rules" of Definition 4.1.*
(3) *The "congruence rules" of Definition 4.17, that is, the rules*

$$\frac{\{\varphi_i \vdash \psi_i \, , \, \psi_i \vdash \varphi_i : i < n\}}{\varpi \varphi_0 \ldots \varphi_{n-1} \vdash \varpi \psi_0 \ldots \psi_{n-1}}$$

for each basic operation symbol ϖ, where n is its arity.

Note that $\mathfrak{G}_{\mathcal{S}}$ is of type ω or ω° depending on whether $\mathcal{S}$ has or has not theorems.

PROPOSITION 4.24. *If S is a selfextensional logic, then $\mathfrak{G}_S$ is adequate for S. If moreover S has the PC then $\mathfrak{G}_S$ is $(\mathbf{t}_\wedge, \mathbf{sq})$-equivalent to $\models_{\mathbf{Alg}\mathfrak{G}_S}$.*

PROOF. The set of sequents $\{\Gamma \vdash \varphi \in \mathrm{Seq}(\mathfrak{G}_S) : \Gamma \vdash_S \varphi\}$, which is the set of axioms of $\mathfrak{G}_S$, is closed under $\vdash_{\mathfrak{G}_S}$: It is closed under the structural rules of (2) because S is a sentential logic, and it is closed under the congruence rules of (3) because S is selfextensional. Thus the sentential logic defined by $\mathfrak{G}_S$ is exactly S, and since we have chosen the type of $\mathfrak{G}_S$ in the right way, $\mathfrak{G}_S$ is adequate for S. Since by definition $\mathfrak{G}_S$ satisfies the congruence rules, we can apply Proposition 4.22 to conclude that $\mathfrak{G}_S$ is $(\mathbf{t}_\wedge, \mathbf{sq})$-equivalent to $\models_{\mathbf{Alg}\mathfrak{G}_S}$.   ⊣

Note that as a consequence, if S is selfextensional and has the PC then the Gentzen system $\mathfrak{G}_S$ satisfies the rules of introduction of Conjunction (4.12) and has the sequents (4.11) as derivable ones; this will simplify some proofs later on.

Observe that in order to prove that $\mathfrak{G}_S$ satisfies all the conditions in Proposition 4.20 it only remains for us to prove that $\mathbf{Alg}\mathfrak{G}_S$ is a variety. We will do this in an indirect way, by seeing that this class of algebras is actually equal to a class already known to be a variety, namely the variety $\mathbf{K}_S$ generated by the Lindenbaum-Tarski algebra of S. Recall that if S is selfextensional, by Proposition 2.43 we know that $\varphi \approx \psi$ holds in $\mathbf{K}_S$ if and only if $\varphi \dashv\vdash_S \psi$; using this, if moreover S has the PC then it is easy to see that the following identities hold in $\mathbf{K}_S$:

$$\varphi \wedge \varphi \approx \varphi \tag{4.13}$$

$$\varphi \wedge \psi \approx \psi \wedge \varphi \tag{4.14}$$

$$\varphi \wedge (\psi \wedge \xi) \approx (\varphi \wedge \psi) \wedge \xi \tag{4.15}$$

Therefore the variety $\mathbf{K}_S$ is *a variety of semilattices with additional structure*; more precisely, it is a variety whose $\wedge$-reducts form a subclass of the variety of semilattices. Our goal is to prove that $\mathbf{K}_S = \mathbf{Alg}\mathfrak{G}_S = \mathbf{Alg}S$. In order to achieve this, we will prove that the Gentzen system $\mathfrak{G}_S$ is $(\mathbf{t}_\wedge, \mathbf{sq})$-equivalent to the equational consequence $\models_{\mathbf{K}_S}$. First note:

LEMMA 4.25. *If S is a selfextensional logic with the PC, then the following hold:*

(1) *An equation $\varphi \approx \psi$ holds in $\mathbf{K}_S$ if and only if $\emptyset \vdash_{\mathfrak{G}_S} \mathbf{sq}(\varphi \approx \psi)$.*
(2) *For any $\Gamma \vdash \varphi \in \mathrm{Seq}(\mathfrak{G}_S)$, $\Gamma \vdash_S \varphi$ (that is, $\emptyset \vdash_{\mathfrak{G}_S} \Gamma \vdash \varphi$) if and only if all equations in $\mathbf{t}_\wedge(\Gamma \vdash \varphi)$ are valid in $\mathbf{K}_S$.*

PROOF. Part (1) is a simple reformulation of Proposition 2.43 in view of Proposition 4.24. Now we prove part (2). If $\Gamma = \emptyset$ and τ is the theorem selected to define $\mathbf{t}_\wedge$, then $\emptyset \vdash_S \varphi$ iff $\tau \vdash_S \varphi$ iff $\tau \approx \varphi$ is valid in $\mathbf{K}_S$, but $\mathbf{t}_\wedge(\emptyset \vdash \varphi) = \tau \approx \varphi$, thus completing the proof in this case. If $\Gamma \neq \emptyset$ and $\Gamma \vdash_S \varphi$, then making repeated use of the PC we obtain $\bigwedge \Gamma \vdash_S (\bigwedge \Gamma) \wedge \varphi$ and also $(\bigwedge \Gamma) \wedge \varphi \vdash_S \bigwedge \Gamma$, that is, $\mathbf{t}_\wedge(\Gamma \vdash \varphi) = (\bigwedge \Gamma) \wedge \varphi \approx \bigwedge \Gamma$ is an axiom of $\mathbf{K}_S$. Conversely, if this last equation is valid in $\mathbf{K}_S$, by 2.43 $\bigwedge \Gamma \vdash_S (\bigwedge \Gamma) \wedge \varphi$ and $(\bigwedge \Gamma) \wedge \varphi \vdash_S \bigwedge \Gamma$; since by the PC we have $\Gamma \vdash_S \bigwedge \Gamma$ and $(\bigwedge \Gamma) \wedge \varphi \vdash_S \varphi$, we get $\Gamma \vdash_S \varphi$. ⊣

PROPOSITION 4.26. *For any selfextensional sentential logic S with the PC, the Gentzen system $\mathfrak{G}_S$ is $(\mathbf{t}_\wedge, \mathbf{sq})$-equivalent to the equational consequence $\models_{\mathbf{K}_S}$.*

PROOF. The proof of condition (Eq2) of Definition 4.14 is trivial: By using equations (4.13) and (4.14) of $\mathbf{K}_S$ it is easy to see that $\varphi \approx \psi \; =\!\|\!\models_{\mathbf{K}_S} \{\varphi \wedge \psi \approx \varphi \,,\, \psi \wedge \varphi \approx \psi\} = \mathbf{t}_\wedge(\mathbf{sq}(\varphi \approx \psi))$, that is, (Eq2). The proof of (Eq1) will also need condition (Eq4):

$$\Gamma \vdash \varphi \; \dashv\!\vdash_{\mathfrak{G}_S} \; \mathbf{sq}(\mathbf{t}_\wedge(\Gamma \vdash \varphi)), \qquad \text{for all } \Gamma \vdash \varphi \in \mathrm{Seq}(\mathfrak{G}_S).$$

To prove this we distinguish between two cases: If $\Gamma = \emptyset$, then we have to show that $\emptyset \vdash \varphi \; \dashv\!\vdash_{\mathfrak{G}_S} \{\tau \vdash \varphi \,,\, \varphi \vdash \tau\}$. By Weakening, $\emptyset \vdash \varphi \vdash_{\mathfrak{G}_S} \tau \vdash \varphi$. Since $\emptyset \vdash_S \tau$, we also have $\varphi \vdash_S \tau$, which implies $\emptyset \vdash_{\mathfrak{G}_S} \varphi \vdash \tau$, and a fortiori $\emptyset \vdash \varphi \vdash_{\mathfrak{G}_S} \varphi \vdash \tau$. On the other hand, using that $\emptyset \vdash_{\mathfrak{G}_S} \emptyset \vdash \tau$ and Cut, we obtain $\{\varphi \vdash \tau, \tau \vdash \varphi\} \vdash_{\mathfrak{G}_S} \emptyset \vdash \varphi$. If $\Gamma \neq \emptyset$ then $\mathbf{sq}(\mathbf{t}_\wedge(\Gamma \vdash \varphi)) = \{(\bigwedge \Gamma) \wedge \varphi \vdash \bigwedge \Gamma, \bigwedge \Gamma \vdash (\bigwedge \Gamma) \wedge \varphi\}$. From the PC for S we get $\emptyset \vdash_{\mathfrak{G}_S} \Gamma \vdash \bigwedge \Gamma$ and $\emptyset \vdash_{\mathfrak{G}_S} (\bigwedge \Gamma) \wedge \varphi \vdash \varphi$, so after several Cuts we obtain $\{(\bigwedge \Gamma) \wedge \varphi \vdash \bigwedge \Gamma, \bigwedge \Gamma \vdash (\bigwedge \Gamma) \wedge \varphi\} \vdash_{\mathfrak{G}_S} \Gamma \vdash \varphi$. For the converse, the PC produces $\emptyset \vdash_{\mathfrak{G}_S} (\bigwedge \Gamma) \wedge \varphi \vdash \bigwedge \Gamma$, which is one half of what we have to prove, and also $\Gamma \vdash \varphi \vdash_{\mathfrak{G}_S} \bigwedge \Gamma \vdash \varphi$; then using the axiom $\bigwedge \Gamma \vdash \bigwedge \Gamma$ and a new Cut we obtain $\Gamma \vdash \varphi \vdash_{\mathfrak{G}_S} \bigwedge \Gamma \vdash (\bigwedge \Gamma) \wedge \varphi$, which completes the proof of (Eq4).

Now we will prove condition (Eq1), that is,

$$\{\Gamma_i \vdash \varphi_i : i \in I\} \vdash_{\mathfrak{G}_S} \Gamma \vdash \varphi \; \Leftrightarrow \; \mathbf{t}_\wedge(\{\Gamma_i \vdash \varphi_i : i \in I\}) \models_{\mathbf{K}_S} \mathbf{t}_\wedge(\Gamma \vdash \varphi).$$

We will first prove ($\Rightarrow$). Assume that $\{\Gamma_i \vdash \varphi_i : i \in I\} \vdash_{\mathfrak{G}_S} \Gamma \vdash \varphi$. In order to prove that $\mathbf{t}_\wedge(\{\Gamma_i \vdash \varphi_i : i \in I\}) \models_{\mathbf{K}_S} \mathbf{t}_\wedge(\Gamma \vdash \varphi)$ it will be enough to take any $A \in \mathbf{K}_S$ and any sequence $\vec{a}$ in A and show that the set of sequents $\Sigma = \{\Gamma \vdash \varphi \in \mathrm{Seq}(\mathfrak{G}_S) : A \models \mathbf{t}_\wedge(\Gamma \vdash \varphi)\,[\vec{a}]\}$ is a theory of $\mathfrak{G}_S$: By Lemma 4.25 it contains all proper axioms of $\mathfrak{G}_S$; note that this also includes the structural axiom $\varphi \vdash \varphi$. Using that $\wedge$ is associative and commutative in every $A \in \mathbf{K}_S$, as we have already mentioned, one can easily prove that Σ is closed under Weakening. Finally it is closed under the Cut rule and the Congruence rules because of the replacement and substitution properties of equality in any algebra.

Now to prove ($\Leftarrow$), if we apply the translation $\mathbf{sq}$ to the right-hand side of (Eq1), by Proposition 4.19 we get $\mathbf{sq}\big(\mathbf{t}_\wedge(\{\Gamma_i \vdash \varphi_i : i \in I\})\big) \mathrel{\vdash\!\!\!\sim}_{\mathfrak{G}_S} \mathbf{sq}\big(\mathbf{t}_\wedge(\Gamma \vdash \varphi)\big)$. But since condition (Eq4) proved before says that every sequent is $\mathfrak{G}_S$-equivalent to its double translation, we obtain exactly the left-hand side of (Eq1). This finishes the proof that $\mathfrak{G}_S$ is $(\mathbf{t}_\wedge, \mathbf{sq})$-equivalent to the equational consequence $\models_{\mathbf{K}_S}$. $\dashv$

We are now ready to obtain the main results of this section.

THEOREM 4.27. *Every selfextensional logic S with the PC has a strongly adequate Gentzen system, namely the system $\mathfrak{G}_S$ defined in 4.23; this Gentzen system is $(\mathbf{t}_\wedge, \mathbf{sq})$-equivalent to $\models_{\mathbf{Alg}S}$, and $\mathbf{Alg}S = \mathbf{Alg}\mathfrak{G}_S$ and they coincide with the variety $\mathbf{K}_S$.*

PROOF. We have seen in Proposition 4.24 that under these assumptions the Gentzen system $\mathfrak{G}_S$ is $(\mathbf{t}_\wedge, \mathbf{sq})$-equivalent to $\models_{\mathbf{Alg}\mathfrak{G}_S}$. Recall that $\mathbf{Alg}\mathfrak{G}_S$ is the class of all algebra reducts of reduced finitary models of $\mathfrak{G}_S$. It has been proved in Rebagliato and Verdú [1995] that in such a case the class $\mathbf{Alg}\mathfrak{G}_S$ is a quasivariety (indeed, *the* equivalent quasivariety semantics for $\mathfrak{G}_S$, uniquely determined by $\mathfrak{G}_S$). In addition, by Proposition 4.26 this Gentzen system is also $(\mathbf{t}_\wedge, \mathbf{sq})$-equivalent to $\models_{\mathbf{K}_S}$. Therefore $\models_{\mathbf{Alg}\mathfrak{G}_S} = \models_{\mathbf{K}_S}$. But $\mathbf{K}_S$ is a variety, hence a quasivariety, and two quasivarieties determining the same equational consequence are equal, that is, $\mathbf{Alg}\mathfrak{G}_S = \mathbf{K}_S$. Therefore $\mathbf{Alg}\mathfrak{G}_S$ is a variety. Hence the three conditions in Proposition 4.20 are satisfied, and we can conclude that $\mathfrak{G}_S$ is strongly adequate for S. As a consequence of this and of Proposition 4.12, $\mathbf{Alg}S = \mathbf{Alg}\mathfrak{G}_S$ and in particular $\mathfrak{G}_S$ is $(\mathbf{t}_\wedge, \mathbf{sq})$-equivalent to $\models_{\mathbf{Alg}S}$. $\dashv$

The presentation of $\mathfrak{G}_S$ given in Definition 4.23 is completely general, and it might not be suitable for practical purposes. However for particular logics more satisfactory presentations are available, as is shown in Chapter 5.

THEOREM 4.28. *If S is a selfextensional sentential logic with the PC then S is strongly selfextensional.*

PROOF. From Theorem 4.27 we know that $\mathfrak{G}_S$ is strongly adequate for S, therefore every full model of S is in particular a model of $\mathfrak{G}_S$. But this Gentzen system satisfies the congruence rules by definition, so every full model of S has the congruence property. This tells us that S is strongly selfextensional. $\dashv$

Thus the open problem mentioned on page 48 has been solved for logics with Conjunction. At this point it may be helpful to summarize some of the preceding results in the following statement:

PROPOSITION 4.29. *Let S be a sentential logic with the PC. Then the follow-ing conditions are equivalent:*

(i) *S is selfextensional.*

(ii) *S is strongly selfextensional.*

(iii) *The Gentzen system $\mathfrak{G}_S$ is strongly adequate for S.*

(iv) *There is a Gentzen system $\mathfrak{G}$ adequate for S that is $(\mathbf{t}, \mathbf{sq})$-equivalent to $\models_{\mathsf{K}}$ for some class $\mathbf{K}$ of algebras and some translation $\mathbf{t}$.*

PROOF. (i)$\Rightarrow$(iii) is contained in Theorem 4.27. The implication (iii)$\Rightarrow$(ii) can be proved in the same way as Theorem 4.28, since in its proof we use just (iii). The implication (ii)$\Rightarrow$(i) is trivial. The implication (i)$\Rightarrow$(iv) is contained in Proposition 4.24, and its converse (iv)$\Rightarrow$(i) is contained in Proposition 4.18. $\dashv$

Note that condition (iv) does not imply that the Gentzen system appearing in it is strongly adequate for S, and thus equal to $\mathfrak{G}_S$; actually the requirements on $\mathfrak{G}$ stated in (iv) are weaker than those in Proposition 4.20.

By examination of the presentation of $\mathfrak{G}_S$ given in Definition 4.23 one sees that the converse of Proposition 2.46 holds for selfextensional logics with the PC, thus obtaining the following characterization of their full models:

COROLLARY 4.30. *Let S be a selfextensional sentential logic with the PC. Then an abstract logic $\mathbb{L}$ is a full model of S if and only if it is a finitary model of S with the congruence property, and having no theorems if S has none.* $\dashv$

In the terminology of Blok and Pigozzi [1989a], a sentential logic is **strongly algebraizable** when it is algebraizable and the equivalent quasivariety semantics is a variety. Using this notion, we have:

PROPOSITION 4.31. *Every selfextensional and algebraizable sentential logic with the PC is strongly algebraizable.*

PROOF. By Proposition 3.2, if S is algebraizable then its equivalent quasivari-ety semantics is precisely **Alg**S. Since S is selfextensional, we can use Theorem 4.27, which says that **Alg**S is a variety. Therefore, S is strongly algebraizable. $\dashv$

Since we already proved in Theorem 3.18 that any Fregean protoalgebraic sen-tential logic with theorems is algebraizable, as a particular case of the preceding result we obtain a radically new proof of a property of Fregean protoalgebraic logics which has been originally obtained by quite different methods[30]:

COROLLARY 4.32 (Czelakowski, Pigozzi). *Every Fregean and protoalgebraic sentential logic with theorems and with the PC is strongly algebraizable.* $\dashv$

---

[30]See Theorem 6.5.5 in Czelakowski [2001a].

We can establish the following parallelism between Theorem 4.27 and Corollary 4.32: While, by the latter, Fregean protoalgebraic logics with the PC and with theorems are algebraizable in the sense of Blok and Pigozzi and the associated class of algebras is a variety, by the former, selfextensional logics with the PC, which form a much wider class and may not be algebraizable in the same sense, determine in a unique way a Gentzen system bearing a very close relationship with them (strong adequacy) and this Gentzen system is algebraizable, in the sense of Rebagliato and Verdú [1993], [1995], with respect to a variety. And in both cases the variety is determined in the same way from the logic itself, it is the variety $\mathbf{K}_S$ characterized by the set of equations $\{\varphi \approx \psi \in Eq : \varphi \dashv\vdash_S \psi\}$.

As a different kind of application of Theorem 4.28, we will show the hereditary character of the Property of Intuitionistic Reductio ad Absurdum (PIRA) dealt with in Section 2.4 (see Definition 2.53). As far as sentential logics are concerned, we can say that S has the PIRA when $\Gamma \vdash_S \neg\varphi$ holds if and only if $\Gamma \cup \{\varphi\}$ is inconsistent (in general, a set is **inconsistent** relative to some closure operator when its closure is the whole universe). We need two properties of such logics:

LEMMA 4.33. *Let S be a sentential logic with the PIRA. Then it satisfies the* **contraposition rule**, *that is, for any $\Gamma \subseteq Fm$ and any $\varphi, \psi \in Fm$, if $\Gamma, \varphi \vdash_S \psi$ then $\Gamma, \neg\psi \vdash_S \neg\varphi$. If moreover S has the PC then for every $\varphi \in Fm$, $\vdash_S \neg(\varphi \wedge \neg\varphi)$, and for every $\varphi, \psi \in Fm$ it holds that $\psi, \neg(\varphi \wedge \psi) \vdash_S \neg\varphi$.*

PROOF. From the PIRA it follows that the set $\{\varphi, \neg\varphi\}$ is always inconsistent, hence any set containing it is also inconsistent. If $\Gamma, \varphi \vdash_S \psi$ then a fortiori we have that $\Gamma, \varphi, \neg\psi \vdash_S \psi$, therefore the set $\Gamma \cup \{\varphi, \neg\psi\}$ is inconsistent, and by the PIRA this implies that $\Gamma, \neg\psi \vdash_S \neg\varphi$. If moreover S has the PC then any formula of the form $\varphi \wedge \neg\varphi$ is inconsistent, so by the PIRA $\neg(\varphi \wedge \neg\varphi)$ is a theorem. Finally, since by the PC the rule $\varphi, \psi \vdash_S \varphi \wedge \psi$ holds, the contraposition rule just proved implies that also $\psi, \neg(\varphi \wedge \psi) \vdash_S \neg\varphi$ holds, as was to be proved.   ⊣

PROPOSITION 4.34. *Let S be a selfextensional sentential logic with the PC and the PIRA. Then every full model of S has the PC and the PIRA.*

PROOF. We already know that every full model of S has the PC. In order to prove that every full model of S has the PIRA it will be enough to prove it for models of the form $\langle A, \mathrm{Fi}_S^A \rangle$, that is, we have to prove that for any $X \cup \{a\} \subseteq A$, $\neg a \in \mathrm{Fi}_S^A(X)$ if and only if $\mathrm{Fi}_S^A(X, a) = A$. If $\neg a \in \mathrm{Fi}_S^A(X)$ then also $\neg a \in \mathrm{Fi}_S^A(X, a)$; since $\varphi, \neg\varphi \vdash_S \psi$, it follows that $\mathrm{Fi}_S^A(X, a) = A$. Conversely, assume that $\mathrm{Fi}_S^A(X, a) = A$. In particular $a \wedge \neg a \in \mathrm{Fi}_S^A(X, a)$. By finitarity and the PC we know that there is some $b \in \mathrm{Fi}_S^A(X)$ such that $a \wedge \neg a \in \mathrm{Fi}_S^A(a, b) = \mathrm{Fi}_S^A(a \wedge b)$: take $b = \neg(a \wedge \neg a)$ if $X = \emptyset$, else $b = a_1 \wedge \cdots \wedge a_k$ for some $a_i \in X$.

Since $a \wedge \neg a$ is inconsistent, it follows that $\mathrm{Fi}_{\mathcal{S}}^{A}(a \wedge b) = \mathrm{Fi}_{\mathcal{S}}^{A}(a \wedge \neg a) = A$. Now, since $\mathcal{S}$ is selfextensional and has the PC, by Theorem 4.28 it is strongly selfextensional, that is, all its full models have the congruence property. In particular for the negation operation we can infer that $\mathrm{Fi}_{\mathcal{S}}^{A}(\neg(a \wedge b)) = \mathrm{Fi}_{\mathcal{S}}^{A}(\neg(a \wedge \neg a)) = \mathrm{Fi}_{\mathcal{S}}^{A}(\emptyset)$. Then, since $\psi, \neg(\varphi \wedge \psi) \vdash_{\mathcal{S}} \neg\varphi$ as proved in Lemma 4.33, we have that $\neg a \in \mathrm{Fi}_{\mathcal{S}}^{A}(b, \neg(a \wedge b)) = \mathrm{Fi}_{\mathcal{S}}^{A}(b) \subseteq \mathrm{Fi}_{\mathcal{S}}^{A}(X)$. This completes the proof that the abstract logic $\langle A, \mathrm{Fi}_{\mathcal{S}}^{A} \rangle$ has the PIRA.   ⊣

PROPOSITION 4.35. *If $\mathcal{S}$ is a sentential logic satisfying the PC with respect to $\wedge$ and the PIRA with respect to $\neg$ and these are the only primitive operations of the formula algebra $\mathbf{Fm}$, then $\mathcal{S}$ is selfextensional and all its full models satisfy the PC, the PIRA and have the congruence property.*

PROOF. In view of Theorem 4.28 and Proposition 4.34 we have only to prove that $\mathcal{S}$ is selfextensional. We have already observed after Definition 2.45 that the PC implies that $\Lambda(\mathcal{S})$ is a congruence with respect to $\wedge$. Now from the Contraposition Rule of Lemma 4.33 we see that from $\varphi \dashv\vdash_{\mathcal{S}} \psi$ it follows that $\neg\varphi \dashv\vdash_{\mathcal{S}} \neg\psi$, which says that $\Lambda(\mathcal{S})$ is a congruence with respect to $\neg$. Since these are all the primitive operations of the algebra, we have proved that $\Lambda(\mathcal{S}) \in \mathrm{Con}\mathcal{S}$, that is, $\mathcal{S}$ is selfextensional.   ⊣

Concerning the relationship between the PC and the DDT see the comments at the end of next section, on page 102.

4.3. Selfextensional logics having the Deduction Theorem

Here we deal with sentential logics satisfying the Deduction-Detachment Theorem, as introduced in Section 2.4. The structure of the section will follow the same pattern as that of Section 4.2: We will first prove a sufficient criterion for a selfextensional logic with the DDT to have a strongly adequate Gentzen system, and then we will introduce a translation and a Gentzen system and prove in several steps that they satisfy the assumptions of the criterion. So we omit many comments[31].

DEFINITION 4.36. *Let $\rightarrow$ be a binary operation symbol, either primitive or defined by a term. Let $\mathfrak{G}$ be a Gentzen system of type ω. Then we say that:*

---

[31] The properties of selfextensional logics with the DDT have been further investigated, with other methods, in Czelakowski and Pigozzi [2004a], [2004b] and in Jansana [2005]. The relationship between the DDT and the property of having a strongly adequate Gentzen system has been dealt with in Font, Jansana, and Pigozzi [2001], [2006].

(1) *The **MP** is a rule of $\mathfrak{G}$ (or that $\mathfrak{G}$ **satisfies the MP**) when for any $\varphi, \psi \in Fm$ and any finite $\Gamma \subseteq Fm$, $\Gamma \vdash \varphi \rightarrow \psi \mathrel{\vdash\joinrel\sim_{\mathfrak{G}}} \Gamma, \varphi \vdash \psi$; and that*

(2) *The **DT** is a rule of $\mathfrak{G}$ (or that $\mathfrak{G}$ **satisfies the DT**) when for any $\varphi, \psi \in Fm$, and any finite $\Gamma \subseteq Fm$, $\Gamma, \varphi \vdash \psi \mathrel{\vdash\joinrel\sim_{\mathfrak{G}}} \Gamma \vdash \varphi \rightarrow \psi$.*

LEMMA 4.37. *If the DT is a rule of a Gentzen system $\mathfrak{G}$, then every finitary model of $\mathfrak{G}$ has the DT. If the MP is a rule of $\mathfrak{G}$ then all its models have the MP. Moreover, if $\mathfrak{G}$ is adequate for a sentential logic $\mathcal{S}$ and $\mathcal{S}$ has the MP, then the MP is a rule of $\mathfrak{G}$ (and thus every model of $\mathfrak{G}$ has it).*

PROOF. Let $\mathbb{L}$ be a finitary model of $\mathfrak{G}$, and assume that $\mathfrak{G}$ satisfies the DT. Then suppose that $b \in C(X, a)$ for some $X \cup \{a, b\} \subseteq A$. There is some finite $X_0 \subseteq X$ such that $b \in C(X_0, a)$, and thus we can find suitable variables $\Gamma_0 \cup \{p, q\} \subseteq Var$ and an homomorphism $h \in \mathrm{Hom}(\boldsymbol{Fm}, \boldsymbol{A})$ such that $h[\Gamma_0] = X_0$, $h(p) = a$ and $h(q) = b$. Since by the DT, $\Gamma_0, p \vdash q \mathrel{\vdash\joinrel\sim_{\mathfrak{G}}} \Gamma_0 \vdash p \rightarrow q$ and $\mathbb{L}$ is a model of $\mathfrak{G}$, we obtain $a \rightarrow b \in C(X_0) \subseteq C(X)$; therefore $\mathbb{L}$ satisfies the DT. Now assume that $\mathfrak{G}$ has the MP; since $\emptyset \mathrel{\vdash\joinrel\sim_{\mathfrak{G}}} \varphi \rightarrow \psi \vdash \varphi \rightarrow \psi$, also $\emptyset \mathrel{\vdash\joinrel\sim_{\mathfrak{G}}} \varphi \rightarrow \psi, \varphi \vdash \psi$. Therefore, any model $\mathbb{L}$ of $\mathfrak{G}$ satisfies $b \in C(a, a \rightarrow b)$ for all $a, b \in A$. As we observed after Definition 2.47, this is enough to guarantee that $\mathbb{L}$ has the MP. Finally, if $\mathfrak{G}$ is adequate for $\mathcal{S}$ and $\mathcal{S}$ has the MP, then $\varphi \rightarrow \psi, \varphi \vdash_{\mathcal{S}} \psi$, so $\emptyset \mathrel{\vdash\joinrel\sim_{\mathfrak{G}}} \varphi \rightarrow \psi, \varphi \vdash \psi$. Then using Cut and Weakening we can show that $\Gamma \vdash \varphi \rightarrow \psi \mathrel{\vdash\joinrel\sim_{\mathfrak{G}}} \Gamma, \varphi \vdash \psi$, that is, the MP is a rule of the Gentzen system $\mathfrak{G}$. ⊣

PROPOSITION 4.38. *Let $\mathcal{S}$ be a sentential logic with the DDT, and let $\mathfrak{G}$ be a Gentzen system that has the DT and such that the following conditions hold:*

(1) *$\mathfrak{G}$ is adequate for $\mathcal{S}$.*

(2) *$\mathfrak{G}$ is $(\mathbf{t}, \mathbf{sq})$-equivalent to $\models_{\mathbf{Alg}\mathfrak{G}}$ for some translation $\mathbf{t}$.*

(3) *$\mathbf{Alg}\mathfrak{G}$ is a variety.*

Then $\mathfrak{G}$ is strongly adequate for $\mathcal{S}$.

PROOF. Since $\mathbf{Alg}\mathfrak{G}$ is a variety, we can apply Proposition 4.18 as in the proof of 4.20 and conclude that $\mathbf{Alg}\mathcal{S} \subseteq \mathbf{Alg}\mathfrak{G}$; but $\mathbf{Alg}\mathfrak{G} \subseteq \mathbf{Alg}\mathcal{S}$ by 4.9, because $\mathfrak{G}$ is adequate for $\mathcal{S}$, therefore $\mathbf{Alg}\mathfrak{G} = \mathbf{Alg}\mathcal{S}$. Now let $\boldsymbol{A} \in \mathbf{Alg}\mathfrak{G}$ and let $\mathbb{L} = \langle \boldsymbol{A}, \mathcal{C} \rangle$ be any finitary and reduced model of $\mathfrak{G}$ over $\boldsymbol{A}$. From 4.18 it follows that $\mathfrak{G}$ satisfies the congruence rules, therefore $\mathbb{L}$ has the congruence property. By assumption $\mathfrak{G}$ has the DT, and it has the MP because by 4.9 it is a model of $\mathcal{S}$; so it has the DDT. Now we can apply Proposition 2.49 to conclude that $\mathbb{L}$ is a full model of $\mathcal{S}$; but since it is reduced we obtain $\mathcal{C} = \mathcal{F}i_{\mathcal{S}}\boldsymbol{A}$. The characterization of Proposition 4.12 tells us that $\mathfrak{G}$ is strongly adequate for $\mathcal{S}$. ⊣

First we present the translation:

DEFINITION 4.39. *For any sentential logic S with the DDT we define the translation* $\mathbf{t}_\rightarrow$ *from* $\mathrm{Seq}(\boldsymbol{Fm})$ *to* $\mathrm{Eq}(\boldsymbol{Fm})$ *as follows:*

$$\mathbf{t}_\rightarrow(\emptyset \vdash \varphi) = \{\varphi \approx p \rightarrow p\}$$

$$\mathbf{t}_\rightarrow(\Sigma \vdash \varphi) = \{\delta_{i_1} \rightarrow (\ldots \rightarrow (\delta_{i_k} \rightarrow \varphi)\ldots) \approx p \rightarrow p\}$$

where $\Sigma = \{\delta_{i_1}, \ldots, \delta_{i_k}\} \neq \emptyset$ *and we assume that* $k \geqslant 1$ *and* $i_1 < \cdots < i_k$ *according to a fixed enumeration of the whole set* Fm; p *is a fixed variable.*

Observe that $\{\delta_1, \ldots, \delta_k\} \vdash_S \varphi$ if and only if $\emptyset \vdash_S \delta_1 \rightarrow (\ldots \rightarrow (\delta_k \rightarrow \varphi)\ldots)$, by the DDT, and that the translation has been so designed in order to obtain that $\mathbf{t}_\rightarrow(\{\delta_1, \ldots, \delta_k\} \vdash \varphi) = \mathbf{t}_\rightarrow(\emptyset \vdash \delta_1 \rightarrow (\ldots \rightarrow (\delta_k \rightarrow \varphi)\ldots))$ (here, assuming the δ_i are already ordered according to a fixed enumeration of Fm).

PROPOSITION 4.40. *Let S be a sentential logic with the DDT and let $\mathfrak{G}$ be a Gentzen system adequate for S such that $\mathfrak{G}$ has the DT and satisfies the congruence rules. Then the Gentzen system $\mathfrak{G}$ is* $(\mathbf{t}_\rightarrow, \mathbf{sq})$-*equivalent to* $\models_{\mathbf{Alg}\mathfrak{G}}$.

PROOF. Note that from the assumptions it follows that every model of $\mathfrak{G}$ has the DDT. We first prove condition (Eq2) of 4.14: $\varphi \approx \psi \dashv\models_{\mathbf{Alg}\mathfrak{G}} \mathbf{t}_\rightarrow(\mathbf{sq}(\varphi \approx \psi))$, that is, $\varphi \approx \psi \dashv\models_{\mathbf{Alg}\mathfrak{G}} \{\varphi \rightarrow \psi \approx p \rightarrow p, \psi \rightarrow \varphi \approx p \rightarrow p\}$. Take any $\boldsymbol{A} \in \mathbf{Alg}\mathfrak{G}$ and let $\mathbb{L} = \langle \boldsymbol{A}, \mathrm{C} \rangle$ be any a reduced finitary model of $\mathfrak{G}$ over $\boldsymbol{A}$: From the assumptions it follows that $\mathbb{L}$ has the congruence property, therefore in this algebra it holds that $a = b$ iff $\mathrm{C}(a) = \mathrm{C}(b)$. Now, for every $a, b \in A$ we see that $a \rightarrow a = b \rightarrow b$, because, by the DDT, we have $\mathrm{C}(a \rightarrow a) = \mathrm{C}(\emptyset) = \mathrm{C}(b \rightarrow b)$. From this it follows $\varphi \approx \psi \models_{\mathbf{Alg}\mathfrak{G}} \{\varphi \rightarrow \psi \approx p \rightarrow p, \psi \rightarrow \varphi \approx p \rightarrow p\}$. To prove the converse, assume that $a \rightarrow b = c \rightarrow c$ and $b \rightarrow a = c \rightarrow c$: Then $\mathrm{C}(a \rightarrow b) = \mathrm{C}(b \rightarrow a) = \mathrm{C}(c \rightarrow c) = \mathrm{C}(\emptyset)$ and by the DDT $\mathrm{C}(a) = \mathrm{C}(b)$, which implies $a = b$. Using this, we obtain that $\{\varphi \rightarrow \psi \approx p \rightarrow p, \psi \rightarrow \varphi \approx p \rightarrow p\} \models_{\mathbf{Alg}\mathfrak{G}} \varphi \approx \psi$. We have proved (Eq2).

To prove condition (Eq1) we must prove that

$$\{\Gamma_i \vdash \varphi_i : i \in I\} \hspace{1pt}\mid\hspace{-5pt}\sim_{\mathfrak{G}} \Gamma \vdash \varphi \Leftrightarrow \mathbf{t}_\rightarrow(\{\Gamma_i \vdash \varphi_i : i \in I\}) \models_{\mathbf{Alg}\mathfrak{G}} \mathbf{t}_\rightarrow(\Gamma \vdash \varphi).$$

($\Rightarrow$) Let $\boldsymbol{A} \in \mathbf{Alg}\mathfrak{G}$, $\mathbb{L} = \langle \boldsymbol{A}, \mathrm{C} \rangle$ a reduced finitary model of $\mathfrak{G}$, and let $\vec{a}$ be a sequence of elements of A such that for each $i \in I$, $\boldsymbol{A} \models \mathbf{t}_\rightarrow(\Gamma_i \vdash \varphi_i) [\vec{a}]$. For a fixed i assume that $\Gamma_i = \{\delta_1, \ldots, \delta_k\} \neq \emptyset$, so we have $(\delta_1 \rightarrow (\ldots \rightarrow (\delta_k \rightarrow \varphi_i)\ldots))^{\boldsymbol{A}}(\vec{a}) = (p \rightarrow p)^{\boldsymbol{A}}(\vec{a}) = p^{\boldsymbol{A}}(\vec{a}) \rightarrow p^{\boldsymbol{A}}(\vec{a})$. Therefore $\mathrm{C}((\delta_1 \rightarrow (\ldots \rightarrow (\delta_k \rightarrow \varphi_i)\ldots))^{\boldsymbol{A}}(\vec{a})) = \mathrm{C}(\emptyset)$ and hence by the DDT $\varphi_i^{\boldsymbol{A}}(\vec{a}) \in \mathrm{C}(\delta_1^{\boldsymbol{A}}(\vec{a}), \ldots, \delta_k^{\boldsymbol{A}}(\vec{a}))$; if $\Gamma_i = \emptyset$ then what we have is $\varphi_i^{\boldsymbol{A}}(\vec{a}) = p^{\boldsymbol{A}}(\vec{a}) \rightarrow p^{\boldsymbol{A}}(\vec{a})$ which implies $\varphi_i^{\boldsymbol{A}}(\vec{a}) \in \mathrm{C}(\emptyset)$. Thus for all $i \in I$ we have $\varphi_i^{\boldsymbol{A}}(\vec{a}) \in \mathrm{C}(\Gamma_i^{\boldsymbol{A}}(\vec{a}))$. Since $\mathbb{L}$ is a model of $\mathfrak{G}$, this implies that $\varphi^{\boldsymbol{A}}(\vec{a}) \in \mathrm{C}(\Gamma^{\boldsymbol{A}}(\vec{a}))$. Now if $\Gamma = \emptyset$ this implies that $C(\varphi^{\boldsymbol{A}}(\vec{a})) = C(\emptyset) = C((p \to p)^{\boldsymbol{A}}(\vec{a}))$; since $\mathbb{L}$ has the congruence property and is reduced, this implies that $\varphi^{\boldsymbol{A}}(\vec{a}) = (p \to p)^{\boldsymbol{A}}(\vec{a})$. If on the other hand $\Gamma = \{\gamma_1, \ldots, \gamma_n\} \neq \emptyset$, then we have $\varphi^{\boldsymbol{A}}(\vec{a}) \in C(\gamma_1^{\boldsymbol{A}}(\vec{a}), \ldots, \gamma_n^{\boldsymbol{A}}(\vec{a}))$ which similarly leads to $(\gamma_1 \to (\ldots \to (\gamma_n \to \varphi) \ldots))^{\boldsymbol{A}}(\vec{a}) = (p \to p)^{\boldsymbol{A}}(\vec{a})$. So in both cases we have obtained that $\boldsymbol{A} \models \mathbf{t}_\to (\Gamma \vdash \varphi) [\vec{a}]$, as was to be proved.

($\Leftarrow$) Let Σ be the closed set of $\vdash_{\mathfrak{G}}$ generated by the set $\{\Gamma_i \vdash \varphi_i : i \in I\}$. By Proposition 4.4 the abstract logic $\mathbb{L}_\Sigma = \langle \boldsymbol{Fm}, C_\Sigma \rangle$ is a finitary model of $\mathfrak{G}$. Therefore by assumption it has the DDT and the congruence property. As a consequence, $\widetilde{\boldsymbol{\Omega}}(\mathbb{L}_\Sigma) = \boldsymbol{\Lambda}(\mathbb{L}_\Sigma) = \{\langle \varphi, \psi \rangle : C_\Sigma(\varphi) = C_\Sigma(\psi)\}$. Now suppose that $\Gamma_i = \{\eta_1, \ldots, \eta_s\} \neq \emptyset$. Since by construction we have that $\varphi_i \in C_\Sigma(\eta_1, \ldots, \eta_s)$, it follows by the DDT that $C_\Sigma(\eta_1 \to (\ldots \to (\eta_s \to \varphi_i) \ldots)) = C_\Sigma(\emptyset) = C_\Sigma(p \to p)$, that is, $\langle \eta_1 \to (\ldots \to (\eta_s \to \varphi) \ldots), p \to p \rangle \in \widetilde{\boldsymbol{\Omega}}(\mathbb{L}_\Sigma)$; this implies that $\boldsymbol{Fm}/\widetilde{\boldsymbol{\Omega}}(\mathbb{L}_\Sigma) \models \mathbf{t}_\to (\Gamma_i \vdash \varphi) [\pi]$ where π is the interpretation defined by the natural projection onto the quotient. If, on the other hand, $\Gamma_i = \emptyset$, then $C_\Sigma(\varphi_i) = C_\Sigma(\emptyset) = C_\Sigma(p \to p)$ which as before implies that $\boldsymbol{Fm}/\widetilde{\boldsymbol{\Omega}}(\mathbb{L}_\Sigma) \models \varphi_i \approx p \to p [\pi]$. Thus for all $i \in I$ we have that $\boldsymbol{Fm}/\widetilde{\boldsymbol{\Omega}}(\mathbb{L}_\Sigma) \models \mathbf{t}_\to (\Gamma_i \vdash \varphi_i) [\pi]$. Since $\boldsymbol{Fm}/\widetilde{\boldsymbol{\Omega}}(\mathbb{L}_\Sigma) \in \mathsf{Alg}\mathfrak{G}$, the assumption of this part implies that $\boldsymbol{Fm}/\widetilde{\boldsymbol{\Omega}}(\mathbb{L}_\Sigma) \models \mathbf{t}_\to (\Gamma \vdash \varphi) [\pi]$. Now a similar process in the opposite direction, distinguishing the cases Γ empty and Γ non-empty, proves that $\varphi \in C_\Sigma(\Gamma)$. Therefore $\{\Gamma_i \vdash \varphi_i : i \in I\} \vdash_{\mathfrak{G}} \Gamma \vdash \varphi$. $\dashv$

Now we present the Gentzen system.

DEFINITION 4.41. *Let S be a selfextensional logic with the DDT. Define a Gentzen system $\mathfrak{G}'_S$ of type ω by the following axioms and rules on* $\mathrm{Seq}(\boldsymbol{Fm})$:

(1) *The "proper axioms" $\Gamma \vdash \varphi$ for all $\Gamma \vdash \varphi \in \mathrm{Seq}(\boldsymbol{Fm})$ such that $\Gamma \vdash_S \varphi$.*

(2) *The "structural rules" of Definition 4.1.*

(3) *The "congruence rules" of Definition 4.17, that is, the rules*

$$\frac{\{\varphi_i \vdash \psi_i, \ \psi_i \vdash \varphi_i : i < n\}}{\varpi\, \varphi_0 \ldots \varphi_{n-1} \vdash \varpi\, \psi_0 \ldots \psi_{n-1}}$$

for each basic operation symbol ϖ, where n is its arity.

(4) *The rule corresponding to the DT:* $\dfrac{\Gamma, \varphi \vdash \psi}{\Gamma \vdash \varphi \to \psi}$.

PROPOSITION 4.42. *If S is a selfextensional logic with the DDT then $\mathfrak{G}'_S$ is adequate for S and is $(\mathbf{t}_\to, \mathbf{sq})$-equivalent to $\models_{\mathsf{Alg}\mathfrak{G}_S}$.*

PROOF. The set of sequents $\{\Gamma \vdash \varphi \in \mathrm{Seq}(\mathfrak{G}'_S) : \Gamma \vdash_S \varphi\}$, which is the set of axioms of $\mathfrak{G}'_S$, is actually its set of theorems, because it is closed under the rules of $\vdash_{\mathfrak{G}'_S}$: It is closed under the structural rules of (2) because S is a sentential logic, it is closed under the congruence rules of (3) because S is selfextensional, and it is closed under the rule of (4) because S satisfies the DT by assumption. Thus the sentential logic defined by $\mathfrak{B}'_S$ is exactly S, and $\mathfrak{B}'_S$ is of type ω just because S has theorems, that is, $\mathfrak{B}'_S$ is adequate for S. Moreover, since by definition $\mathfrak{B}'_S$ satisfies the congruence rules and the DT, we can apply Proposition 4.40 to conclude that $\mathfrak{B}'_S$ is $(\mathbf{t}_{\rightarrow}, \mathrm{sq})$-equivalent to $\models_{\mathsf{Alg}\,\mathfrak{B}_S}$. $\dashv$

Consider the variety $\mathbf{K}_S$ generated by the Lindenbaum-Tarski algebra $\boldsymbol{Fm}^*$ of S. If S is selfextensional, Proposition 2.43 states that an equation $\varphi \approx \psi$ holds in $\mathbf{K}_S$ if and only if $\varphi \dashv\vdash_S \psi$. If moreover S has the DDT then one can easily prove that the following equations hold in $\mathbf{K}_S$:

$$\varphi \rightarrow \varphi \approx \psi \rightarrow \psi \tag{4.16}$$

$$(\varphi \rightarrow \varphi) \rightarrow \varphi \approx \varphi \tag{4.17}$$

$$\varphi \rightarrow (\psi \rightarrow \xi) \approx (\varphi \rightarrow \psi) \rightarrow (\varphi \rightarrow \xi) \tag{4.18}$$

$$(\varphi \rightarrow \psi) \rightarrow \big((\psi \rightarrow \varphi) \rightarrow \psi\big) \approx (\psi \rightarrow \varphi) \rightarrow \big((\varphi \rightarrow \psi) \rightarrow \varphi\big) \tag{4.19}$$

and from them we recognize that $\mathbf{K}_S$ is a *variety of Hilbert algebras with additional structure* (more precisely, $\mathbf{K}_S$ is a variety such that the class of all its $\rightarrow$-reducts is a subclass of the variety of all Hilbert algebras). Hilbert algebras, studied mainly in Diego [1965], [1966], are also called *positive implication algebras* in the literature, and are the algebraic counterpart of the *logic of positive implication*, the implicative fragment of intuitionistic logic, which is characterized by the Deduction Theorem. They can be equationally defined by the above equations, see Diego [1966] Theorem 3, although they are usually presented with a constant 1 which is the interpretation of the term $p \rightarrow p$, which is an algebraic constant as equation (4.16) shows. Among their properties we highlight the following:

$$\text{If } a \rightarrow b = 1 \text{ and } b \rightarrow a = 1 \text{ then } a = b \tag{4.20}$$

$$a \rightarrow 1 = 1 \tag{4.21}$$

$$\text{If } 1 \rightarrow a = 1 \text{ then } a = 1 \tag{4.22}$$

Alternative presentations of Hilbert algebras, more details and further references can be found in Rasiowa [1974] Section II.2.

LEMMA 4.43. *Let S be a selfextensional logic with the DDT. Then the following hold:*

(1) *An equation $\varphi \approx \psi$ holds in $\mathbf{K}_S$ if and only if $\emptyset \mathrel{\vdash\!\!\!\sim}_{\mathfrak{B}'_S} \mathrm{sq}(\varphi \approx \psi)$.*

(2) *For any $\Gamma \vdash \varphi \in \mathrm{Seq}(\boldsymbol{Fm})$, $\Gamma \vdash_S \varphi$ (that is, $\emptyset \mathrel{\vdash\!\!\!\sim}_{\mathfrak{B}'_S} \Gamma \vdash \varphi$), if and only if $\mathbf{t}_{\rightarrow}(\Gamma \vdash \varphi)$ is an equation valid in $\mathbf{K}_S$.*

PROOF. In view of Proposition 4.42, part (1) is just a reformulation of Proposition 2.43. Let us prove part (2). $\Gamma \vdash_S \varphi$ if and only if either $\emptyset \vdash_S \delta_1 \to (\dots \to (\delta_k \to \varphi) \dots)$, when $\Gamma = \{\delta_1, \dots, \delta_k\}$, or simply $\emptyset \vdash_S \varphi$ otherwise. Since $\emptyset \vdash_S p \to p$, the first fact is equivalent to $\delta_1 \to (\dots \to (\delta_k \to \varphi) \dots) \dashv\vdash_S p \to p$ and the second one is equivalent to $\varphi \dashv\vdash_S p \to p$. Therefore, (2) holds. $\dashv$

PROPOSITION 4.44. *Let S be a selfextensional logic with the DDT. Then the Gentzen system $\mathfrak{G}'_S$ is $(\mathbf{t}_\to, \mathbf{sq})$-equivalent to $\models_{\mathbf{K}_S}$.*

PROOF. Since $\mathbf{t}_\to\big(\mathbf{sq}(\varphi \approx \psi)\big) = \{\varphi \to \psi \approx p \to p, \psi \to \varphi \approx p \to p\}$, condition (Eq2) of Definition 4.14 becomes

$$\{\varphi \to \psi \approx p \to p, \psi \to \varphi \approx p \to p\} \dashv\models_{\mathbf{K}_S} \varphi \approx \psi;$$

the entailment from right to left follows from equation (4.16), while the entailment from left to right follows from property (4.20). Condition (Eq4) is also easy to check: For $\Gamma = \emptyset$, since $\mathbf{sq}\big(\mathbf{t}_\to(\emptyset \vdash \varphi)\big) = \{\varphi \vdash p \to p, p \to p \vdash \varphi\}$, we have to check that

$$\{\varphi \vdash p \to p, p \to p \vdash \varphi\} \dashv\vdash_{\mathfrak{G}'_S} \emptyset \vdash \varphi.$$

From $p \to p \vdash \varphi$ and $\emptyset \vdash p \to p$, an axiom of $\mathfrak{G}'_S$, we obtain $\emptyset \vdash \varphi$ after a Cut. Conversely from $\emptyset \vdash \varphi$, by Weakening we obtain $p \to p \vdash \varphi$, and from the same axiom plus Weakening we derive $\varphi \vdash p \to p$. The case $\Gamma \neq \emptyset$ can be reduced to the case $\Gamma = \emptyset$ because if $\Gamma = \{\delta_1, \dots, \delta_k\}$ then by using the DT rule of $\mathfrak{G}'_S$ we have that $\Gamma \vdash \varphi \dashv\vdash_{\mathfrak{G}'_S} \emptyset \vdash \delta_1 \to (\dots \to (\delta_k \to \varphi) \dots) \dashv\vdash_{\mathfrak{G}'_S} \mathbf{sq}\big(\mathbf{t}_\to(\emptyset \vdash \delta_1 \to (\dots \to (\delta_k \to \varphi) \dots))\big) = \mathbf{sq}\big(\mathbf{t}_\to(\Gamma \vdash \varphi)\big)$.

Now we will prove condition (Eq1), that is,

$$\{\Gamma_i \vdash \varphi_i : i \in I\} \vdash_{\mathfrak{G}'_S} \Gamma \vdash \varphi \;\Leftrightarrow\; \mathbf{t}_\to\big(\{\Gamma_i \vdash \varphi_i : i \in I\}\big) \models_{\mathbf{K}_S} \mathbf{t}_\to(\Gamma \vdash \varphi).$$

($\Rightarrow$): Assume that $\{\Gamma_i \vdash \varphi_i : i \in I\} \vdash_{\mathfrak{G}'_S} \Gamma \vdash \varphi$. In order to prove that $\mathbf{t}_\to\big(\{\Gamma_i \vdash \varphi_i : i \in I\}\big) \models_{\mathbf{K}_S} \mathbf{t}_\to(\Gamma \vdash \varphi)$ it will be enough to take any $A \in \mathbf{K}_S$ and any sequence $\vec{a}$ in A and show that the set of sequents $\Sigma = \{\Gamma \vdash \varphi \in \mathrm{Seq}(\boldsymbol{Fm}) : A \models \mathbf{t}_\to(\Gamma \vdash \varphi)\,[\vec{a}]\}$ is a theory of $\mathfrak{G}'_S$: By Lemma 4.43 it contains all proper axioms of $\mathfrak{G}'_S$; note that this also includes the structural one $\varphi \vdash \varphi$. Using equation (4.21) we have that $\varphi \approx p \to p \models_{\mathbf{K}_S} \psi \to \varphi \approx p \to p$, and this shows that Σ is closed under Weakening. Using equation (4.22) we see that $\{\varphi \approx p \to p, \varphi \to \psi \approx p \to p\} \models_{\mathbf{K}_S} \psi \approx p \to p$, and from this it follows that Σ is closed under the Cut rule. That Σ is closed under the congruence rules follows from replacement for equality together with property (4.20). Finally Σ is trivially closed under the DT rule, because by definition $\mathbf{t}_\to(\Gamma, \varphi \vdash \psi) = \mathbf{t}_\to(\Gamma \vdash \varphi \to \psi)$.

($\Leftarrow$): Since the $\mathbf{t}_\rightarrow$-translation of a set of sequents is a set of equations, if we have $\mathbf{t}_\rightarrow(\{\Gamma_i \vdash \varphi_i : i \in I\}) \models_{\mathbf{K}_S} \mathbf{t}_\rightarrow(\Gamma \vdash \varphi)$, then by Proposition 4.19 we also have $\mathbf{sq}(\mathbf{t}_\rightarrow(\{\Gamma_i \vdash \varphi_i : i \in I\})) \vdash_{\mathfrak{G}'_S} \mathbf{sq}(\mathbf{t}_\rightarrow(\Gamma \vdash \varphi))$, and then by (Eq4) we obtain $\{\Gamma_i \vdash \varphi_i : i \in I\} \vdash_{\mathfrak{G}'_S} \Gamma \vdash \varphi$. ⊣

Now we obtain our main results.

THEOREM 4.45. *Every selfextensional logic S with the DDT has a strongly adequate Gentzen system, namely the system $\mathfrak{G}'_S$ defined in 4.41; this Gentzen system is $(\mathbf{t}_\rightarrow, \mathbf{sq})$-equivalent to $\models_{\mathbf{Alg}S}$; and $\mathbf{Alg}S = \mathbf{Alg}\mathfrak{G}'_S = \mathbf{K}_S$, the variety generated by the Lindenbaum-Tarski algebra of S.*

PROOF. We have seen in Proposition 4.42 that under these assumptions the Gentzen system $\mathfrak{G}'_S$ is $(\mathbf{t}_\rightarrow, \mathbf{sq})$-equivalent to $\models_{\mathbf{Alg}\mathfrak{G}'_S}$. Recall that $\mathbf{Alg}\mathfrak{G}'_S$ is the class of all algebra reducts of reduced finitary models of $\mathfrak{G}'_S$. It has been proved in Rebagliato and Verdú [1995] that in such a case the class $\mathbf{Alg}\mathfrak{G}'_S$ is a quasivariety (indeed, *the* equivalent quasivariety semantics for $\mathfrak{G}'_S$, uniquely determined by $\mathfrak{G}'_S$). By Proposition 4.44 this Gentzen system is also $(\mathbf{t}_\rightarrow, \mathbf{sq})$-equivalent to $\models_{\mathbf{K}_S}$. Therefore by (Eq3), $\models_{\mathbf{Alg}\mathfrak{G}_S} = \models_{\mathbf{K}_S}$. But $\mathbf{K}_S$ is a variety, hence a quasivariety, and two quasivarieties determining the same equational consequence are equal, hence $\mathbf{Alg}\mathfrak{G}'_S = \mathbf{K}_S$, therefore $\mathbf{Alg}\mathfrak{G}'_S$ is a variety. Since by 4.42 $\mathfrak{G}'_S$ is adequate for S, and it has the DT as a rule, we can apply Proposition 4.38 and conclude that $\mathfrak{G}'_S$ is strongly adequate for S. As a consequence, $\mathbf{Alg}S = \mathbf{Alg}\mathfrak{G}'_S$. Therefore $\mathfrak{G}'_S$ is $(\mathbf{t}_\rightarrow, \mathbf{sq})$-equivalent to $\models_{\mathbf{Alg}S}$. ⊣

THEOREM 4.46. *Every selfextensional logic with the DDT is strongly selfextensional.*

PROOF. We know that, by the preceding theorem, all the full models of S will be models of $\mathfrak{G}'_S$. Since this Gentzen system satisfies the congruence rules by definition, all the full models of S will also have the congruence property, that is, S will be strongly selfextensional. ⊣

Thus the open problem mentioned on page 48 has been solved for logics with the Deduction Theorem. Now we summarize some of the preceding results in the following statement:

PROPOSITION 4.47. *Let S be a sentential logic with the DDT. Then the following conditions are equivalent:*

(i) *S is selfextensional.*

(ii) *S is strongly selfextensional.*

(iii) *The Gentzen system $\mathfrak{G}'_S$ is strongly adequate for S.*

(iv) *There is a Gentzen system $\mathfrak{G}$ adequate for S that is $(\mathbf{t}, \mathbf{sq})$-equivalent to $\models_{\mathbf{K}}$ for some class $\mathbf{K}$ of algebras and some translation $\mathbf{t}$.*

PROOF. (i)$\Rightarrow$(iii) is contained in Theorem 4.45. The implication (iii)$\Rightarrow$(ii) can be proved in the same way as Theorem 4.46, since in its proof we use just (iii). The implication (ii)$\Rightarrow$(i) is trivial. The implication (i)$\Rightarrow$(iv) is contained in Proposition 4.42, and its converse (iv)$\Rightarrow$(i) is contained in Proposition 4.18. $\dashv$

Note that condition (iv) does not imply that the Gentzen system appearing in it is strongly adequate for S, and thus equal to $\mathfrak{G}'_S$; actually the requirements on $\mathfrak{G}$ stated in (iv) are weaker than those in Proposition 4.38; for instance (iv) does not require $\mathfrak{G}$ to satisfy the DT rule.

Taking Proposition 2.48 into account we see that the converse of Proposition 2.49 holds, and we get the following characterization of the full models of the logics treated in this section:

COROLLARY 4.48. *Let S be a selfextensional logic with the DDT, and let $\mathbb{L}$ be any abstract logic. Then $\mathbb{L}$ is a full model of S if and only if it is a finitary model of S with the DT and having the congruence property.* $\dashv$

As an application of these constructions we obtain an important property of the Fregean logics with the DDT, parallel to that obtained by Pigozzi and Czelakowski for Fregean protoalgebraic logics having the PC (see our Corollary 4.32); note that here it is not necessary to explicitly assume protoalgebraicity since it follows from the DDT.

PROPOSITION 4.49. *Every selfextensional algebraizable logic with the DDT is strongly algebraizable. In particular, every Fregean logic with the DDT is strongly algebraizable.*

PROOF. Since S is algebraizable, by Proposition 3.2 its equivalent quasivariety semantics is $\mathbf{Alg}S$. Since S is selfextensional, by Theorem 4.45 $\mathbf{Alg}S$ is a variety. Thus S is strongly algebraizable. Now assume that S is Fregean and has the DDT. The latter property implies that S has theorems, and also that S is protoalgebraic (actually, protoalgebraic logics are characterized by a weaker type of Deduction-Detachment Theorem, see Czelakowski and Dziobiak [1991]). Thus S is Fregean, protoalgebraic, and has theorems, and we can apply Theorem 3.18 to conclude that it is regularly algebraizable, and as in the first part we obtain that it is also strongly algebraizable. $\dashv$

Finally, consider what happens with a selfextensional logic S that satisfies both the PC (with respect to $\wedge$) and the DDT (with respect to $\rightarrow$): By Theorem 4.27 the Gentzen system $\mathfrak{G}_S$ defined in 4.23 is strongly adequate for S; but by Theorem

4.45 the same is true for the system $\mathfrak{G}'_S$ of 4.41. Since a strongly adequate Gentzen system, if it exists, is unique, we conclude that both systems are the same (i.e., as consequence relations among sequents), and after comparing them we obtain the (maybe surprising) conclusion that the DT is actually a *derived rule* of the Gentzen system $\mathfrak{G}_S$.

CHAPTER 5

APPLICATIONS TO PARTICULAR
SENTENTIAL LOGICS

In this chapter we determine the classes of $\mathcal{S}$-algebras and of full models for several logics, especially for some which do not fit into the classical approaches to the algebraization of logic. We classify them according to several of the criteria we have been considering, i.e., the properties of the Leibniz, Tarski and Frege operators, which determine the classes of selfextensional logics, Fregean logics, strongly selfextensional logics, protoalgebraic logics, etc. We also study the counterexamples promised in the preceding chapters of this monograph.

It goes without saying that the number of cases we have examined is limited, and that many more are waiting to be studied[32]. In our view this is an interesting program, especially for non-algebraizable logics. Among those already proven in Blok and Pigozzi [1989a] not to be algebraizable we find many quasi-normal and other modal logics like Lewis' S1, S2 and S3, entailment system E, several purely implicational logics like BCI, the system $R_{\rightarrow}$ of relevant implication, the "pure entailment" system $E_{\rightarrow}$, the implicative fragment $S5_{\rightarrow}$ of the Wajsberg-style version of S5, etc. Other non-algebraizable logics not treated in the present monograph are Da Costa's paraconsistent logics C_n (see Lewin, Mikenberg, and Schwarze [1991]), and the "logic of paradox" of Priest [1979] (see Pynko [1995]). This program is also interesting for some algebraizable logics whose class of $\mathcal{S}$-algebras is already known, but whose full models have not yet been investigated; this includes Łukasiewicz many-valued logics (see Rodríguez, Torrens, and Verdú [1990]), BCK logic and some of its neighbours (see Blok and Pigozzi [1989a] Theorem 5.10), the equivalential fragments of classical and intuitionistic logics

---

[32]The full models of several subintuitionistic logics have been determined in Bou [2001]; those of certain positive modal logics have been studied in Jansana [2002]; those of the version of Łukasiewicz logic that preserves degrees of truth, in Font, Gil, Torrens, and Verdú [2006]; and, more in general, those of any logic preserving degrees of truth with respect to a variety of residuated lattices (see Galatos, Jipsen, Kowalski, and Ono [2007]) are determined in Bou, Esteva, Font, Gil, Godo, Torrens, and Verdú [2009]. Most of these logics are non-protoalgebraic.

(see Blok and Pigozzi [1989a] Section 5.2.6), Rasiowa's logic with semi-negation (see her [1988] p. 391), Nelson's logic of constructive falsity (see Wójcicki [1988] Section 5.3), etc.

In general, for protoalgebraic logics the class of S-algebras will be determined by using Proposition 3.2. But for these logics what is really interesting and new is the determination of their full models. To this end, for protoalgebraic logics and also for non-protoalgebraic logics, we will usually use the equivalence between conditions (i) and (iii) of Proposition 2.21. In order to apply this result we will first determine, usually by ad-hoc arguments, the S-filters on S-algebras, and we will then use for each particular logic a particular theorem, let us call it "Theorem T" here, which already exists in the literature and does not refer to full models. Theorem T is similar to Proposition 2.21 in that it states, for an arbitrary abstract logic, the equivalence between three conditions (i), (ii) and (iii), having the same form as those in 2.21: its condition (i) contains some characterization of the abstract logic, while its condition (iii) states the existence of a bilogical morphism between the arbitrary abstract logic and an abstract logic of a particular kind, which after the ad-hoc characterizations we have mentioned is recognized to consist of an S-algebra and all the S-filters on it. Thus, applying Proposition 2.21, we conclude that the full models of S are the abstract logics characterized as in part (i) of Theorem T. Moreover, the particular theorems of this kind that we will consider have the peculiarity that their condition (ii) uses the Frege relation instead of the Tarski congruence, and includes explicitly that it is a congruence. Therefore, the logics for which we characterize the full models by using the method just described are strongly selfextensional.

However, to be able to find the full models of S from characterizations of the S-filters on S-algebras one often needs to use a Hilbert-style axiomatization of S, and this is not always at hand. For logics defined by a Gentzen system an alternative and more direct way exists, exemplified in Section 5.1. One starts with a Gentzen system $\mathfrak{G}$ which is adequate for the logic; then one finds the $\mathfrak{G}$-algebras, proves that they form a variety, and that $\mathfrak{G}$ is $(\mathbf{t}, \mathbf{sq})$-equivalent to $\models_{\mathbf{Alg}\mathfrak{G}}$, with $\mathbf{t} = \mathbf{t}_\wedge$ or $\mathbf{t} = \mathbf{t}_\rightarrow$ (in this second case $\mathfrak{G}$ must have the DT as a rule). Then Propositions 4.20 or 4.38 tell us that this $\mathfrak{G}$ is *the* Gentzen system strongly adequate for S; thus the full models of S can be described from the models of $\mathfrak{G}$, and moreover $\mathbf{Alg}S = \mathbf{Alg}\mathfrak{G}$. Of course this only works for logics satisfying the assumptions in the proposition applied, that is, for selfextensional logics with either the PC or the DDT, which is the case of most of the logics treated in the literature in this way.

A detailed account of how to apply existing results in order to follow either of the ways just summarized will only be given for the case of the conjunction-disjunction fragment of classical logic (Section 5.1.1). In other examples we just mention the properties concerning the particular logic which are relevant here, and refer the reader to the literature; otherwise this chapter would become excessively long.

Let us mention that, once the $\mathcal{S}$-algebras and the full models of a particular sentential logic $\mathcal{S}$ have been identified, then all "isomorphism theorems" proven separately in each case in Font and Verdú [1991], Jansana [1995], Rebagliato and Verdú [1993], Rius [1992], Rodríguez [1990] and Verdú [1986] become particular instances of our Theorem 2.30, which states that for each algebra A there is an isomorphism between the lattices of all full models of $\mathcal{S}$ over A and of all congruences of A which give an $\mathcal{S}$-algebra in the quotient.

5.1. Some non-protoalgebraic logics

While in the protoalgebraic cases the determination of the class of $\mathcal{S}$-algebras is covered by a general result (Proposition 3.2) which confirms that it is the class already obtained by the matrix approach, this is not the case for non-protoalgebraic logics, where we do not have an alternative theory. Generally speaking, in each case one has to confirm by ad-hoc arguments that the class of $\mathcal{S}$-algebras is the class one hopes to find (or, maybe, in some cases, that it is not !). However, all the cases reviewed in this section are selfextensional and satisfy the PC, and thus we will determine **Alg**$\mathcal{S}$ and **FMod**$\mathcal{S}$ using the results of Section 4.2, since a strongly adequate Gentzen system is available. It is also interesting to note that in all the cases in this section it has been found that **Alg**$^*\mathcal{S}$ is a proper subclass of **Alg**$\mathcal{S}$; this is not, however, a characteristic of non-protoalgebraic logics: The logic WR discussed in Section 5.4.1 is not protoalgebraic either but its two classes of algebras are equal; this logic, however, is discussed later on because it has an algebraizable "strong version", and the algebraic analysis of it helps in the analysis of WR and conversely.

5.1.1. $\text{CPC}_{\wedge\vee}$, the $\{\wedge,\vee\}$-fragment of Classical Logic

This sentential logic may be considered a paradigmatic example of the usefulness of our approach precisely due to its simplicity: it can be defined by a very natural Gentzen system (see below), but also semantically by the single matrix $\langle \mathbf{2}, \{1\} \rangle$ where $\mathbf{2} = \langle \{0,1\}, \wedge, \vee \rangle$ is the two-element distributive lattice, which generates the whole variety **D** of *distributive lattices*. So this logic is determined by a single algebra. The variety **D** it generates is also generated by the Linden-baum-Tarski algebra of the logic. While these are natural associations between $CPC_{\wedge\vee}$ and **D**, in Font and Verdú [1991] deeper connections are established, which with a small adjustment can be used to prove that, in effect, the distributive lattices are the $CPC_{\wedge\vee}$-algebras, and to determine the full models of $CPC_{\wedge\vee}$.

It is proved in Font and Verdú [1991] Proposition 2.8 that $CPC_{\wedge\vee}$ is not pro-toalgebraic. The class of algebra reducts of its reduced matrices is determined in Font, Guzmán, and Verdú [1991]: it is the class of distributive lattices with maximum 1 such that if $a < b$ there is a c with $a \vee c \neq 1$ and $b \vee c = 1$; this condi-tion is dual to the so-called "Wallman disjunction property" (see Birkhoff [1973]), and the distributive lattices that satisfy it form a proper subclass of **D** that is not even a quasivariety; a surprising fact, which follows from Corollary 3.6 of Cig-noli [1991], is that its finite members are the finite Boolean algebras. It seems clear from the beginning that this class is not the algebraic counterpart of $CPC_{\wedge\vee}$.

The logic $CPC_{\wedge\vee}$ is Fregean, as are all two-valued logics (see page 68), hence it is also selfextensional. Moreover, using that **2** is at the same time the generator of **D** and the support of the single matrix used to define $CPC_{\wedge\vee}$, it is trivial to check that $\varphi \dashv\vdash_{CPC_{\wedge\vee}} \psi$ if and only if $\mathbf{D} \models \varphi \approx \psi$. Therefore by Proposition 2.43 we conclude that $\mathbf{K}_{CPC_{\wedge\vee}} = \mathbf{D}$. Now the determination of $\mathbf{Alg}CPC_{\wedge\vee}$ is straightforward: Since $CPC_{\wedge\vee}$ satisfies the PC and is selfextensional, we can apply Theorem 4.27 and conclude that $\mathbf{Alg}CPC_{\wedge\vee} = \mathbf{K}_{CPC_{\wedge\vee}} = \mathbf{D}$. As for the full models of $CPC_{\wedge\vee}$, we will illustrate in detail the two ways of determining them mentioned before.

(a) Using the Hilbert-style presentation of $CPC_{\wedge\vee}$ given in Dyrda and Prucnal [1980] one easily proves that on a distributive lattice, the $CPC_{\wedge\vee}$-filters are the filters of the lattice plus the empty set (note that $CPC_{\wedge\vee}$ does not have theorems). Since $\mathbf{D} = \mathbf{Alg}CPC_{\wedge\vee}$, by Proposition 2.21 the full models of $CPC_{\wedge\vee}$ are abstract logics $\mathbb{L} = \langle A, \mathcal{C} \rangle$ such that there is a bilogical morphism between them and the abstract logics constituted by a distributive lattice and the closure system of all its lattice filters plus the empty set. Now we will show that these are all the abstract logics $\mathbb{L} = \langle A, C \rangle$ such that:

   (1) $\mathbb{L}$ is finitary.
   (2) $\mathbb{L}$ satisfies the PC and the PDI.
   (3) $\mathbb{L}$ does not have theorems, that is, it satisfies $C(\emptyset) = \emptyset$.

In Font and Verdú [1991] a very close class of abstract logics is studied, namely those satisfying (1) and (2) but, instead of (3), the condition

   (3') $C(\emptyset) = \bigcap \{ T \in \mathcal{C} : T \neq \emptyset \}$, that is, $\mathbb{L}$ is non-pseudoaxiomatic.

Such abstract logics are called *distributive*, and the following result having the form of 2.21 is proved in Theorem 4.2 of Font and Verdú [1991]: An abstract logic is distributive if and only if there is a bilogical morphism between it and the abstract logic determined by all lattice filters of a distributive lattice. However, distributive abstract logics are not exactly the full models of $\mathrm{CPC}_{\wedge\vee}$: While the empty set is always a closed set of every full model of $\mathrm{CPC}_{\wedge\vee}$, it may not be a closed set of every distributive logic; for instance there are distributive lattices with maximum 1, where $\{1\}$ is the least filter of the lattice. However, it is easy to check that everything works equally smoothly after replacing (3') with (3), and Theorem 4.2 of Font and Verdú [1991] can be reproduced in our case, with the addition of $\emptyset$ to the filters of the lattice. As a consequence of all this and of Proposition 2.21 we conclude that, in effect, the full models of $\mathrm{CPC}_{\wedge\vee}$ are the abstract logics satisfying conditions (1), (2) and (3) above.

Of course this procedure depends on "guessing" the three properties just mentioned that will eventually characterize $\mathrm{CPC}_{\wedge\vee}$; this guess can probably be guided by the results on Gentzen systems we consider next.

(b) The second way is more direct, and does not even need the previous proof that $\mathbf{Alg}\mathrm{CPC}_{\wedge\vee} = \mathbf{D}$. Consider the Gentzen system $\mathfrak{G}_D$ presented in Font and Verdú [1991]: it is of type ω°, and has the structural rules and the following rules corresponding to the PC and the PDI:

$$(\wedge\vdash) \quad \frac{\Gamma,\varphi,\psi\vdash\xi}{\Gamma,\varphi\wedge\psi\vdash\xi} \qquad\qquad (\vdash\wedge) \quad \frac{\Gamma\vdash\varphi \quad \Gamma\vdash\psi}{\Gamma\vdash\varphi\wedge\psi}$$

$$(\vee\vdash) \quad \frac{\Gamma,\varphi\vdash\xi \quad \Gamma,\psi\vdash\xi}{\Gamma,\varphi\vee\psi\vdash\xi} \qquad (\vdash\vee) \quad \frac{\Gamma\vdash\varphi}{\Gamma\vdash\varphi\vee\psi} \quad \frac{\Gamma\vdash\psi}{\Gamma\vdash\varphi\vee\psi}$$

Proposition 2.4 of Font and Verdú [1991] proves that, in our terminology, $\mathfrak{G}_D$ is adequate for $\mathrm{CPC}_{\wedge\vee}$. In Theorem 4.9 of Rebagliato and Verdú [1993], where this Gentzen system is called $\mathcal{G}_3$, it is proved that $\mathfrak{G}_D$ is $(\mathbf{t}_\wedge, \mathbf{sq})$-equivalent to $\models_\mathbf{D}$, and in Corollary 4.5 of Font and Verdú [1991] it is proved that $\mathbf{D}$ is the class of all algebra reducts of the reduced models of $\mathfrak{G}_D$, that is, that $\mathbf{Alg}\mathfrak{G}_D = \mathbf{D}$, which is a variety. Now we can use our Proposition 4.20 and conclude that $\mathfrak{G}_D$ is strongly adequate for $\mathrm{CPC}_{\wedge\vee}$, and our Proposition 4.12 implies that $\mathbf{D} = \mathbf{Alg}\mathrm{CPC}_{\wedge\vee}$; by inspection of the rules of $\mathfrak{G}_D$ we see that the full models of $\mathrm{CPC}_{\wedge\vee}$, which are the finitary models of $\mathfrak{G}_D$ without theorems, are the abstract logics satisfying (1), (2) and (3) above.

By Theorem 4.28, the logic $\mathrm{CPC}_{\wedge\vee}$ is strongly selfextensional, and we know that it is Fregean. However, since it is neither protoalgebraic, nor pseudo-axiomatic, our Propositions 3.15 and 3.16 concerning the relation between theories and full models on the formula algebra do not apply to it. Actually, for every non-empty $\Gamma \in Th\mathrm{CPC}_{\wedge\vee}$, the abstract logic $\mathrm{CPC}_{\wedge\vee}^{\Gamma}$ is not a full model of $\mathrm{CPC}_{\wedge\vee}$ precisely because it has theorems, but it is straightforward to check that it satisfies the PC and the PDI, and it is obviously finitary, so it only lacks condition (3) to be a full model of $\mathrm{CPC}_{\wedge\vee}$; and just adding the empty theory to it makes it a full model, as proved in Proposition 4.11 of Font and Verdú [1991], where we also see that the mapping $\Gamma \mapsto (\mathrm{CPC}_{\wedge\vee}^{\Gamma})_{\emptyset}$, using the notation introduced in page 62, is an order-preserving embedding of $Th\mathrm{CPC}_{\wedge\vee}$ into $\mathcal{FMod}_{\mathrm{CPC}_{\wedge\vee}} \boldsymbol{Fm}$.

Finally let us mention that, as shown in Font and Verdú [1991], the non-linear four-element distributive lattice, equipped with a closure system whose closed sets are just $\{1\}$ and the universe, provides an example of a finitary model of $\mathrm{CPC}_{\wedge\vee}$ that is not a full model of it, thus confirming that the converse of part (1) of Proposition 2.9 is not true, and that in general arbitrary models of a logic may not inherit its main metalogical properties, like the PDI in this case (and hence the congruence property, by Corollary 4.30).

5.1.2. The logic of lattices

In the last part of Rebagliato and Verdú [1993] a Gentzen system related to *the variety* **Lat** *of lattices* is considered. Let us call it $\mathfrak{G}_{L}$; it is defined by the structural rules, the rules $(\wedge \vdash)$ and $(\vdash \wedge)$ and $(\vdash \vee)$ of the previous section, and the weakened form of $(\vee \vdash)$ with $\Gamma = \emptyset$. It is proved there that the sentential logic $\mathcal{G}_{L}$ defined by this calculus is non-protoalgebraic, that $\mathbf{Alg}\mathfrak{G}_{L} = \mathbf{Lat}$, and that $\mathfrak{G}_{L}$ is $(\mathbf{t}_{\wedge}, \mathbf{sq})$-equivalent to $\models_{\mathbf{Lat}}$. Again, our Propositions 4.12 and 4.20 and Theorem 4.28 imply that $\mathcal{G}_{L}$ is strongly selfextensional, that $\mathfrak{G}_{L}$ is strongly adequate for it, that the $\mathcal{G}_{L}$-algebras are all lattices, and the full models of $\mathcal{G}_{L}$ are all the finitary abstract logics without theorems satisfying the PC and the following weakening of the PDI:

$$\forall a, b \in A, \; \mathrm{C}(a \vee b) = \mathrm{C}(a) \cap \mathrm{C}(b) \qquad\qquad \text{(WPDI)}$$

Thus not only is the Gentzen system $\mathfrak{G}_{L}$ naturally associated with the variety of lattices in the sense of Rebagliato and Verdú [1993], but the sentential logic $\mathcal{G}_{L}$ defined by $\mathfrak{G}_{L}$ is also naturally associated with the variety of lattices in the sense of our theory; and we did not need a Hilbert-style presentation of the logic to prove it. Thus the sentential logic $\mathcal{G}_{L}$ deserves to be called *the logic of lattices*; note that in Rebagliato and Verdú [1993] it is also proved that the variety of lattices cannot be the equivalent algebraic semantics of any algebraizable logic, in the sense of Blok and Pigozzi [1989a].

We now prove that $\mathcal{G}_L$ is not Fregean, thus offering a quite natural and simple example of a strongly selfextensional but non-Fregean logic. We reason by contradiction, and assume that any axiomatic extension of $\mathcal{G}_L$ has the property of congruence with respect to $\vee$. Let $\varphi, \psi, \xi \in Fm$; the PC implies that $\varphi, \psi \dashv\vdash_{\mathcal{G}_L} \varphi, \varphi \wedge \psi$ and that $\varphi, \xi \dashv\vdash_{\mathcal{G}_L} \varphi, \varphi \wedge \xi$, that is, that $\langle \psi, \psi \wedge \varphi \rangle \in \Lambda_{\mathcal{G}_L}(\varphi)$ and $\langle \xi, \varphi \wedge \xi \rangle \in \Lambda_{\mathcal{G}_L}(\varphi)$. From this, by our assumption it follows that $\langle \psi \vee \xi, (\varphi \wedge \psi) \vee (\varphi \wedge \xi) \rangle \in \Lambda_{\mathcal{G}_L}(\varphi)$, that is, $\varphi, \psi \vee \xi \dashv\vdash_{\mathcal{G}_L} \varphi, (\varphi \wedge \psi) \vee (\varphi \wedge \xi)$, and by using the PC we obtain that $\varphi \wedge (\psi \vee \xi) \vdash_{\mathcal{G}_L} (\varphi \wedge \psi) \vee (\varphi \wedge \xi)$. But the PC and the WPDI together imply that $(\varphi \wedge \psi) \vee (\varphi \wedge \xi) \dashv\vdash_{\mathcal{G}_L} \varphi \wedge (\psi \vee \xi)$. Therefore we have proved that $\mathrm{Cn}_{\mathcal{G}_L}(\varphi, \psi \vee \xi) = \mathrm{Cn}_{\mathcal{G}_L}((\varphi \wedge \psi) \vee (\varphi \wedge \xi)) = \mathrm{Cn}_{\mathcal{G}_L}(\varphi, \psi) \cap \mathrm{Cn}_{\mathcal{G}_L}(\varphi, \xi)$. Using finitarity and the PC this easily implies that for any $\Gamma \subseteq Fm$, $\mathrm{Cn}_{\mathcal{G}_L}(\Gamma, \psi \vee \xi) = \mathrm{Cn}_{\mathcal{G}_L}(\Gamma, \psi) \cap \mathrm{Cn}_{\mathcal{G}_L}(\Gamma, \xi)$, that is, that $\mathcal{G}_L$ satisfies the PDI. But this would imply that $\mathcal{G}_L = \mathrm{CPC}_{\wedge\vee}$, which is certainly not the case because $\mathsf{Alg}\mathcal{G}_L = \mathsf{Lat}$ while $\mathsf{Alg}\mathrm{CPC}_{\wedge\vee} = \mathsf{D}$. Therefore $\mathcal{G}_L$ cannot be Fregean.

5.1.3. Belnap's four-valued logic, and other related logics

Belnap's four-valued logic[33] was introduced as an independent sentential logic in Belnap [1977] (see also Anderson, Belnap, and Dunn [1992] Section 81), and it corresponds to the system of *tautological entailments* or *first-degree entailments* of Anderson and Belnap [1975]. Let us call it here DM, because this sentential logic, whose language has $\wedge, \vee, \neg$ as connectives, is determined by the four-element De Morgan lattice M_4, which generates *the variety of all De Morgan lattices*. The original definition does not use M_4 as a matrix, but as a generalized matrix; actually, the consequence relation $\vdash_{\mathrm{DM}}$ is defined using the ordering relation of M_4, and essentially, it amounts to saying that $\varphi_1, \ldots, \varphi_n \vdash_{\mathrm{DM}} \psi$ if and only if for any $h \in \mathrm{Hom}(Fm, M_4)$, $h(\varphi_1) \wedge \cdots \wedge h(\varphi_n) \leqslant h(\psi)$.

This case is fairly similar to $\mathrm{CPC}_{\wedge\vee}$, except that it is not Fregean. It was treated with the techniques of abstract logics in Font and Verdú [1988], [1989a] and again in Font [1997], more thoroughly in the last case. It is proved that DM is not protoalgebraic, is selfextensional but not Fregean, and has the PC; therefore we can conclude that it is strongly selfextensional. The DM-algebras are the De Morgan lattices while $\mathsf{Alg}^*\mathrm{DM}$ is a proper subclass, and the full models of DM have been determined in Font [1997]; actually they already appear in Font and Verdú [1988], where they are called *De Morgan logics*. In Font [1997] the following Gentzen system is presented: it is of type ω°, and in addition to structural rules it has

---

[33]Also known in the literature as "Dunn-Belnap's four valued logic", and very often denoted as $\mathcal{FOUR}$; see Dunn [1976] and Dunn and Restall [2002].

the rules for the system presented above for $CPC_{\wedge\vee}$ plus three rules involving negation:

$$\frac{\Gamma \vdash \varphi}{\Gamma \vdash \neg\neg\varphi} \qquad \frac{\Gamma, \varphi \vdash \psi}{\Gamma, \neg\neg\varphi \vdash \psi} \qquad \frac{\varphi \vdash \psi}{\neg\psi \vdash \neg\varphi}$$

It is proved that this system is strongly adequate for DM, and thus the full models of DM are the finitary abstract logics without theorems satisfying the PC, the PDI, the Property of Double Negation: $C(a) = C(\neg\neg a)$, and the Property of Weak Contraposition: $a \in C(b)$ implies $\neg b \in C(\neg a)$. It has been proved in Font [1997] that the full models of DM can be characterized as those finitary abstract logics without theorems whose closure system $\mathcal{C}$ has a basis made of $\vee$-prime $\wedge$-filters that is closed under the mapping $\Phi(X) = \{y \in A : \neg y \notin X\}$ (where $X \subseteq A$) and such that Φ is idempotent on that basis (this mapping is a re-definition of the one used in the representation of De Morgan algebras and lattices, see Balbes and Dwinger [1974]). Note that to use the results in Font and Verdú [1988] the condition involving the empty set (i.e., that the abstract logics under consideration do not have theorems) must be explicitly added, as in Section 5.1.1.

Observe that in particular from the above results we get the characterization (which is essentially already in Anderson and Belnap [1975]) that Belnap's logic is the weakest sentential logic without theorems satisfying the PC, the PDI and the Properties of Double Negation and Weak Contraposition. A similar characterization has been obtained in Pynko [1995b], but with the De Morgan Laws in the place of the Weak Contraposition. Note, however, that this is not a "best" characterization of the sentential logic, in the sense that it does not characterize its full models; actually, De Morgan Laws are weaker than Weak Contraposition.

Several *extensions* are considered in Font [1997]. If we add a nullary connective (i.e., a constant) $\top$ to the language, interpret it as the maximum of M_4, and add it as an axiom to DM we find a logic whose $\mathcal{S}$-algebras are *De Morgan algebras* (bounded De Morgan lattices). By extending the Gentzen system for DM to all sequents and adding the axiom $\emptyset \vdash \top$ to it one obtains similar results; the full models of this extension are like the full models of DM but with the condition $C(\emptyset) = \emptyset$ replaced by the condition $C(\top) = C(\emptyset)$. Dually, one can add a constant $\bot$ interpreted as the minimum of M_4, and acting as an inconsistent element; the results are essentially the same.

A different kind of extension is the logic K_3, whose $\mathcal{S}$-algebras are *Kleene lattices*, a proper subvariety of De Morgan lattices. It can be obtained from DM by adding one more rule to its Gentzen system, namely the axiom $\varphi \wedge \neg\varphi \vdash \psi \vee \neg\psi$; it is the implication-less fragment of Kleene's strong three-valued logic. Semantically, it is defined from the three-element Kleene lattice M_3 in the same way as DM is defined from M_4, through the ordering relation. Note that M_3 is a De Morgan lattice, and generates the variety **K3** of Kleene lattices. The full models of this logic are the full models of DM that satisfy the above mentioned rule, i.e., such that $a \vee \neg a \in C(b \wedge \neg b)$ for all $a, b \in A$. Combining with the addition of $\top$ or $\bot$ just mentioned we find a logic whose S-algebras are exactly *Kleene algebras*, and whose full models are the abstract logics satisfying all the just mentioned additional properties together.

5.1.4. The implication-less fragment of IPC and its extensions

This logic, denoted as IPC* and whose language is $(\wedge, \vee, \neg)$, is shown in Blok and Pigozzi [1989a] to be non-protoalgebraic, and to have an *algebraic semantics* in the precise sense of this paper, namely the class **PCDL** of *pseudo-complemented distributive lattices*, (see Balbes and Dwinger [1974] Chapter VIII for the history and basic theory of these structures). This logic is studied in Rebagliato and Verdú [1993] from the point of view of the algebraization of Gentzen systems. There it is proved that **PCDL** cannot be the equivalent algebraic semantics of any algebraizable logic, and a Gentzen system of type ω° is presented. Since IPC* has theorems, to match the results of Rebagliato and Verdú [1993] with our approach we must modify this system by allowing the empty set to appear in the left part of its sequents, that is, we consider it as defined on the whole set of sequents Seq($\boldsymbol{Fm}$); let us call this modified Gentzen system $\mathfrak{G}_P$. It is not difficult to check that the following results of Rebagliato and Verdú [1993] still hold for it: $\mathfrak{G}_P$ is adequate for IPC*, **PCDL** is the class of algebraic reducts of the reduced models of $\mathfrak{G}_P$, that is, **Alg**$\mathfrak{G}_P$ = **PCDL**, the finitary models of $\mathfrak{G}_P$ are the finitary abstract logics satisfying the PC, the PDI and the PIRA, and $\mathfrak{G}_P$ is $(\mathsf{t}_\wedge, \mathsf{sq})$-equivalent to $\models_{\textbf{PCDL}}$. Since IPC* satisfies the PC, and **PCDL** is a variety, we can use our Proposition 4.20 to conclude that $\mathfrak{G}_P$ is strongly adequate for IPC* and that the full models of IPC* are the finitary abstract logics satisfying the PC, the PDI and the PIRA; and we can also use Proposition 4.12 to conclude that **Alg**IPC* = **PCDL**, and Theorem 4.28 to conclude that IPC* is strongly selfextensional. It is easy to check that the properties PC, PDI and PIRA are preserved under axiomatic extensions, and that they imply the congruence property; therefore we conclude that IPC* is Fregean. Finally let us mention that Theorem 3.15 of Rebagliato and Verdú [1993] proves that **Alg***IPC* is the proper subclass of **PCDL** containing the algebras in this class such that for any a, b with ab there is a $c \neq 1$ such that $a \leqslant c$ and $\neg(\neg a \wedge b) \leqslant c \vee b$, where $1 = \neg(a \wedge \neg a)$ is the maximum of the algebra. So again this is a case where the ordinary theory of matrices does not lead us to the class of algebras naturally associated with the logic.

It is easy to see that completely analogous results can be obtained for the denumerable chain of extensions of IPC^* dealt with in Rebagliato and Verdú [1993]. They are all the sentential logics which as abstract logics are the full models of IPC^*; they correspond to all the subvarieties of **PCDL**. These logics are nonprotoalgebraic (this is shown in Rebagliato and Verdú [1993]), strongly selfextensional and Fregean because they are axiomatic extensions of IPC^*. In Rebagliato and Verdú [1993] Gentzen systems for all of these logics are presented, and it is proved that for each one of them the Gentzen system is $(\mathbf{t}_\wedge, \mathbf{sq})$-equivalent to $\models_\mathbf{V}$ for the corresponding variety $\mathbf{V} \subseteq \mathbf{PCDL}$. Although it is not explicitly worked out in Rebagliato and Verdú [1993], it is straightforward to see that the finitary models of the Gentzen system are the full models of IPC^* that satisfy, in addition, a condition that is the abstract counterpart of the additional axiom for the Gentzen system, and that the algebraic reducts of the reduced full models are precisely the algebras in the corresponding subvariety $\mathbf{V}$. Thus, by Proposition 4.20 each one of these Gentzen systems is strongly adequate for its sentential logic, and, by Proposition 4.12, the class of $\mathcal{S}$-algebras for this logic is the subvariety $\mathbf{V}$.

5.2. Some Fregean algebraizable logics

It results from Theorem 3.18 and Proposition 3.19 that Fregean algebraizable logics are regularly algebraizable and strongly selfextensional. Since any logic being an extension of an algebraizable one is also algebraizable (with the same defining equations and equivalence formulas, see Blok and Pigozzi [1989a] Corollary 4.9), and any logic being an axiomatic extension of a Fregean one is also Fregean, this group includes every axiomatic extension of each of its members; the best-known of them are $IPC_\rightarrow$, the implicative fragment of intuitionistic propositional calculus IPC, sometimes called *logic of positive implication*, as well as any other fragment provided it contains implication, IPC itself, and all their axiomatic extensions, including classical logic CPC.

The examples we review here all belong to the class of logics studied in Rasiowa [1974]; it is proved in Blok and Pigozzi [1989a] that all such logics are algebraizable with equivalence formulas $\{p \rightarrow q , q \rightarrow p\}$ and defining equation $p \approx p \rightarrow p$. By Proposition 3.2 the class of $\mathcal{S}$-algebras of these logics is their equivalent quasivariety semantics. All these algebras have an algebraic constant 1, which interprets $p \rightarrow p$, such that $\langle \mathbf{A}, \{1\} \rangle$ is a reduced matrix for $\mathcal{S}$. All our examples can be formalized with axioms and Modus Ponens as the sole rule of inference, thus the S-filters are the so-called *implicative filters*: subsets $F \subseteq A$ such that $1 \in F$ and are closed under Modus Ponens: if $a \to b \in F$ and $a \in F$ then $b \in F$. By Corollary 3.11, the full models on S-algebras are just the families of implicative filters that contain a fixed implicative filter. Once the classes of S-algebras and of full models of one of these logics are known, the S-algebras for all its axiomatic extensions are obtained by adding the equation $\varphi \approx p \to p$ for each proper axiom φ, and the class of full models is obtained by adding the condition $h(\varphi) \in C(\emptyset)$ for all $h \in \mathrm{Hom}(\boldsymbol{Fm}, \boldsymbol{A})$, for each proper axiom φ. While this yields a "standard" procedure, in some cases nicer characterizations of the classes of full models have already been obtained. A summary of the properties of some cases follows:

$S = \mathrm{IPC}_\to$: the implicative fragment of the intuitionistic propositional logic. It is well-known that the $\mathrm{IPC}_\to$-algebras are the *Hilbert algebras* (see page 99). An abstract logic $\mathbb{L} = \langle A, C \rangle$ is a full model of $\mathrm{IPC}_\to$ iff it is finitary and satisfies the DDT or Deduction Theorem, see Verdú [1978] II.3.3. These abstract logics are the finitary models of the Gentzen system that has the structural rules and DT and MP as proper rules; so this Gentzen system is strongly adequate for $\mathrm{IPC}_\to$. As a consequence of the results of Section 4.3, it is the only Gentzen system with that property, and it is $(\mathbf{t}_\to, \mathbf{sq})$-equivalent to $\models_\mathbf{H}$, where $\mathbf{H}$ is the variety of Hilbert algebras. In Section 5.2.1 we mention other Gentzen systems which are adequate but not strongly adequate for $\mathrm{IPC}_\to$.

$S = \mathrm{CPC}_\to$: the implicative fragment of classical propositional logic. This is the axiomatic extension of $\mathrm{IPC}_\to$ obtained by taking $\big((\varphi \to \psi) \to \varphi \big) \to \varphi$, commonly known as *Peirce's Law*, as additional axiom. The $\mathrm{CPC}_\to$-algebras are the *implication algebras*, see Rasiowa [1974] IX.7.1. From Theorem 3 of Verdú [1987] it follows that an abstract logic $\mathbb{L} = \langle A, C \rangle$ is a full model of $\mathrm{CPC}_\to$ if and only if it is finitary, satisfies the DDT, and $\big((a \to b) \to a \big) \to a \in C(\emptyset)$ for all $a, b \in A$; this last condition can be substituted by the condition that the closure system has a basis of maximal sets. A semantical characterization is that $\mathbb{L}$ is projectively generated from the implicative reduct of the two-element Boolean algebra by the set of all homomorphisms which map some designated set into $\{1\}$.

$S = \mathrm{IPC}^+$: the fragment of IPC without negation, sometimes also called *positive logic*. By Theorem X.2.1 of Rasiowa [1974], the IPC^+-algebras are the *relatively pseudo-complemented lattices*, and by Theorem II.4.1 of Verdú

[1978], the full models of IPC^+ can be characterized as those finitary abstract logics satisfying the DDT, the PC and the WPDI; this last one can be replaced by the full PDI.

$\mathcal{S} = IPC$: the intuitionistic propositional logic. The IPC-algebras are the *Heyting algebras* (also called pseudo-Boolean algebras), and by Theorem 2.6 in Font and Verdú [1989b], the full models of IPC can be characterized as the finitary abstract logics satisfying the DDT, the PC, the WPDI or the PDI, and an additional condition which can be either the existence of an inconsistent element, if we include the falsum $\perp$ but not negation in the similarity type, or the PIRA, if we put negation but not $\perp$ in the similarity type.

$\mathcal{S} = CPC$: the classical propositional logic. Naturally, the class of the CPC-algebras is the class of *Boolean algebras*, and depending on the similarity type chosen to present them we have different characterizations of the full models of CPC: For $(\neg, \vee)$ it is already in Theorem 3 of Bloom and Brown [1973]: Finitary, with the PDI and the PRA. For $(\neg, \rightarrow)$ it appears in Theorem II.5.6 of Verdú [1978]: Finitary, the DDT and the PRA. For $(\neg, \wedge)$ it appears in Theorem 13 of Verdú [1979]: Finitary, the PC and the PRA. Also from Theorem 9 of Verdú [1985] it follows that we can formulate it with only $\rightarrow$: Finitary, the DDT, with a closure system $\mathcal{C}$ having a basis of maximal closed sets, and with an inconsistent element. Of course, if one wants all the usual connectives to be primitive, then the corresponding conditions must be simultaneously present.

The observations on the Gentzen system strongly adequate for $\mathcal{S}$ that we made in the case of $IPC_{\rightarrow}$ can also be reproduced for all the logics in this section. In each case the conditions on C characterizing the full models produce the necessary rules for $\mathfrak{G}$; see Wójcicki [1988] pp. 116 ff. for a discussion on the expression of the PIRA and the PRA as Gentzen-style rules. Note that for fragments of IPC these conditions agree with the properties used in Porębska and Wroński [1975] to characterize them. Here we have explicitly mentioned the fragments with implication already studied in the literature, but the other fragments (which are non-protoalgebraic) also admit these kinds of characterization, as detailed in Sections 5.1.1 and 5.1.4; see also Bloom [1977] for the fragments with Conjunction.

5.2.1. Alternative Gentzen systems adequate for $IPC_{\rightarrow}$ not having the full Deduction Theorem

Since $IPC_{\rightarrow}$ satisfies the DDT, it follows from the results in Section 4.3 that there is one and only one Gentzen system of type ω whose finitary models are exactly the full models of $IPC_{\rightarrow}$; as we have already noted, these are characterized

as those finitary abstract logics satisfying the DDT. Here we present a denumerable chain of Gentzen systems, all adequate for IPC$_\rightarrow$, but none of them strongly
adequate for it. Consider the following Gentzen-style rules, where $n \in \omega$:

$$\text{(MP)} \quad \frac{\Gamma \vdash \varphi \quad \Gamma, \psi \vdash \xi}{\Gamma, \varphi \rightarrow \psi \vdash \xi} \qquad\qquad \text{(DT}n) \quad \frac{\Gamma, \varphi \vdash \psi}{\Gamma \vdash \varphi \rightarrow \psi} \quad \text{if card}(\Gamma) \leqslant n$$

Strictly speaking, (DTn) is the abbreviated formulation of a set of $n + 1$ explicit
Gentzen-style rules. Call $\mathfrak{G}_n$ the Gentzen system of type ω defined by the Structural Rules of Definition 4.1 and the rules (MP) and (DTn). This sequence of
Gentzen systems has been studied in García Lapresta [1991][34]; it is obviously
increasing, because (DT$n + 1$) includes (DTn), and, as we shall see, they are
all different. For all $n \geqslant 2$ the sentential logic defined by $\mathfrak{G}_n$ is exactly IPC$_\rightarrow$
(while it is not so for $n = 0, 1$; these two last cases are dealt with in Section 5.4.4).
However, neither of them is strongly adequate for it, since the models of $\mathfrak{G}_n$ are
exactly the abstract logics satisfying (MP) and the abstract version of (DTn). We
have the following characterization in the line of Proposition 2.21: An abstract
logic $\mathbb{L} = \langle A, \mathcal{C} \rangle$ is a model of $\mathfrak{G}_n$, for $n \geqslant 2$, iff there is a bilogical morphism
between it and an abstract logic $\mathbb{L}' = \langle A', \mathcal{C}' \rangle$ where A' is a Hilbert algebra and
$\mathcal{C}'$ is a family of implicative filters containing all those generated by at most $n + 1$
elements. This enables us to find examples of models of $\mathfrak{G}_n$ which are not models of $\mathfrak{G}_{n+1}$ (indeed, they can be found on a finite Hilbert algebra, so they are all
finitary). As a consequence, we see that (DTn) does not imply (DT$n + 1$), thus
$\mathfrak{G}_n$ is strictly weaker than $\mathfrak{G}_{n+1}$. This is an example of an algebraic proof of a
proof-theoretic fact. Since we know that the full models of IPC$_\rightarrow$ satisfy the full
DT and their reduction must consist of a Hilbert algebra and all its implicative filters, the above results imply that these Gentzen systems are not strongly adequate
for IPC$_\rightarrow$.

5.3. Some modal logics

In the vast domain of modal logics, we will refer in detail only to those already
studied with the techniques of abstract logics; this has been done in Font and
Verdú [1989b], Jansana [1992], [1995], after the early attempts of Font [1980],
Font and Verdú [1979]. The algebraizability and equivalential character of many
quasi-normal and quasi-classical modal logics is also analyzed in Czelakowski

---

[34]See also Bou, Font, and García Lapresta [2004], where further results around these Gentzen
systems and the logics they define are presented.

[2001a] Sections 3.4–3.6 . In this section we consider modal formulas and algebras as having some set of non-modal connectives plus a unary connective $\square$ intended to represent the necessity operator.

Many modal logics, understood as sentential logics in the technical sense we have given to this term (i.e., as consequence relations rather than as sets of theorems), come in pairs, one normal and one quasi-normal. In Blok and Pigozzi [1989a] 5.2.1 it is pointed out that in the literature there are several ways of defining a given modal logic (namely S5), which generate the same theorems but define different consequences; the difference lies in the *Rule of Necessitation*, which can be taken in its *strong* form ($\varphi \vdash \square\varphi$) or in its *weak* or *restricted* form ($\vdash \varphi$ implies $\vdash \square\varphi$); see also our page 57.

This situation is very general. We denote by S and S_N the pairs of the weak and the corresponding strong version of a normal modal logic[35]; both S and S_N have the same theorems (the formulas of the "system" of modal logic, as it is usually called), and S has the MP as the only rule of inference, while S_N has in addition the strong Rule of Necessitation; note that S satisfies the restricted form of the Rule of Necessitation. One can prove that S_N is algebraizable while S is not (unless $p \to \square p$ is a theorem, which would imply $S_N = S$), and that S is protoalgebraic and selfextensional, while S_N is not selfextensional. It follows from Proposition 4.5 of Jansana [1995] that for any algebra A, $\widetilde{\Omega}_A(\mathcal{F}i_S A) = \widetilde{\Omega}_A(\mathcal{F}i_{S_N} A)$, and as a consequence $\mathbf{Alg}S = \mathbf{Alg}S_N$; that is, both logics have the same associated class of algebras. For the smallest normal modal logic K we find the class of *normal modal algebras*; for KT, Lemmon's *extension algebras*; for S4, Tarski's *closure algebras*, also called *topological Boolean algebras*; and for S5, Halmos' *monadic Boolean algebras*. Other axiomatic extensions of K generate the corresponding classes of algebras in the way explained at the beginning of Section 5.2.

Since $\mathbf{Alg}S = \mathbf{Alg}S_N$, the algebraization of the two logics differs in the relationship between the sentential logic and the class of algebras established in the Completeness Theorem 2.22, that is, they differ in their associated abstract logics rather than in their associated algebras. This is a case where the need for the determination of the full models of the logics is clear; at present we have found a strongly adequate Gentzen system only for S, thus characterizing the full models of S, while the full models of S_N seem to resist such characterizations, and are

---

[35]The denominations of *local* and *global* (instead of those of "weak" and "strong") for the logics denoted here by S and S_N have become widespread in the literature, see for instance Kracht [2007]. These terms originated in the relational semantics for these modal logics: In the best behaved cases, the two logics of each pair are complete with respect to the same class of frames, one as its local consequence and the other as is global consequence.

determined only as the strong versions of the full models of S, in the way we explain below.

All the logics considered can be axiomatized by some set of axioms and just Modus Ponens and Necessitation (weak or strong) as the sole rules. Thus, the S_N-filters on the S_N-algebras (which form a subclass of normal modal algebras) are all the open filters (i.e., all Boolean filters F such that $\Box[F] \subseteq F$), regardless of the properties of the unary operator $\Box$, since these (besides the Rule of Necessitation) are expressed by equating the axioms to 1, and 1 belongs to every filter. On the other hand, the S-filters on S-algebras are all the Boolean filters, since the weak Rule of Necessitation is automatically satisfied by the axiomatization of the algebras (precisely, by the condition $\Box 1 = 1$). Then Proposition 2.21 together with several results in Font and Verdú [1989b] and Jansana [1995] give characterizations of the full models of S. To describe them we need some specific notations:

If A is an algebra of suitable type (which includes the unary operation $\Box$), then we denote by A^- the $\Box$-less reduct of A; and for any an abstract logic $\mathbb{L} = \langle A, C \rangle$, we put $\mathbb{L}^- = \langle A^-, C \rangle$ and call this its *non-modal reduct*. Finally, if C is a closure system, we consider the closure system C^+ of its *open sets*, that is, $C^+ = \{T \in C : \Box[T] \subseteq T\}$, and its associated closure operator C^+; then for any $\mathbb{L} = \langle A, C \rangle$ we consider its associated *strong version* $\mathbb{L}^S = \langle A, C^+ \rangle$. For sentential logics we have that $(S)^S = S_N$.

We can then prove that an abstract logic $\mathbb{L} = \langle A, C \rangle$ is a full model of S if and only if $\mathbb{L}^-$ is a full model of CPC (i.e., it is finitary and satisfies the DDT and the PRA, for instance) and the operator C satisfies one or more properties directly coming from the modal axioms of the particular S; for instance, for K it is the condition that $\Box[C(X)] \subseteq C(\Box[X])$ for all $X \subseteq A$, for KT one adds $C(X) \subseteq C(\Box[X])$, for K4 one adds $\Box[C(\Box[X])] \subseteq C(\Box[X])$. For S4 (=KT4) it is enough to put the two last conditions together, but full models of S4 can be more compactly characterized by the condition $C^+ = C \circ \Box$, and also by saying that the mapping $X \mapsto C(\Box[X])$ is a closure operator, see Font and Verdú [1989b] Definition 3.1 and Proposition 3.2.

For S4 and S5, the paper Font and Verdú [1989b] contains the following nice characterizations, assuming that we have all the operations $\wedge, \vee, \rightarrow, \neg$ in the type. We define the following new operations: $a \vee^+ b = \Box a \vee \Box b$, $a \rightarrow^+ b = \Box a \rightarrow b$ and $\neg^+ a = \neg \Box a$, and we put $A^+ = \langle A, \wedge, \vee^+, \rightarrow^+, \neg^+ \rangle$ and $\mathbb{L}^+ = \langle A^+, C^+ \rangle$. It has been proved that $\mathbb{L}$ is a full model of S4 iff $\mathbb{L}^-$ is a full model of CPC, $\mathbb{L}^+$ is a full model of IPC, and they have the same theorems, and that $\mathbb{L}$ is a full model of S5 iff both $\mathbb{L}^-$ and $\mathbb{L}^+$ are full models of CPC and have the same theorems. These results are the abstract expression of a deeper fact, made apparent also in other studies of these logics: that the modal part of S4 is "intuitionistic" in character, while that of S5 is "classical"; for a detailed discussion of this phenomenon for these two logics and for their intuitionistic counterparts, see Font and Verdú [1989b] and Font and Verdú [1990].

Concerning full models of the normal versions, for the time being we can only say that an abstract logic $\mathbb{L}$ is a full model of $\mathcal{S}_N$ iff there is another abstract logic $\mathbb{L}_0$ that is a full model of $\mathcal{S}$ and such that $\mathbb{L} = (\mathbb{L}_0)^S$; in this situation, we can prove that there is only one such $\mathbb{L}_0$.

A similar study, along the lines of the preceding paragraphs, is done in Jansana [1992] for the well-known logic GL of provability. There it is proved that the specific modal condition for full models of GL is that if $a \in \mathrm{C}(\square[X] \cup X, \square a)$ then $\square a \in \mathrm{C}(\square[X])$.

In Font and Verdú [1989b], Jansana [1995], *modal logics with an intuitionistic base* are also considered. Everything works as in the classical case, except that the non-modal reduct $\mathbb{L}^-$ of $\mathbb{L}$ must now be a full model of IPC instead of CPC. Some partial results on interior operators on implicative structures in Font [1980], Font and Verdú [1979] seem to indicate that it is possible to further weaken the non-modal reduct of the logics to other fragments of IPC, and similar results can be obtained.

Finally, in Jansana [1995] two denumerable chains of extensions of K, one between K and K4 and the other between K and K4B, are considered; the full models of the weak versions also admit characterizations similar to the one given before for S4 using $\mathbb{L}^+$, but with a more elaborate definition of the reduct A^+ and of the closure system $\mathcal{C}^+$.

The overall conclusion of this section is that a large class of modal logics, on a classical or a non-classical base, can be treated with parallel procedures; they are those whose non-modal part is algebraizable, and whose modal part contains at least the axiom for K and the Rule of Necessitation in its weak or strong form. It would be an interesting task to examine weaker modal logics, in particular those which have received some algebraic treatment, like those studied in Lemmon [1966], and also the classical and quasi-classical logics presented in Chellas [1980] (where the algebraic models are introduced through exercises) and in Blok and Köhler [1983]. The classical ones are clearly algebraizable, as it is easy to see that they belong to Rasiowa's group; instead of the Rule of Necessitation they have the weaker rule $\varphi \leftrightarrow \psi \vdash \square\varphi \leftrightarrow \square\psi$, which can also be taken in a strong and in a weak sense.

5.3.1. A logic without a strongly adequate Gentzen system

We will describe a simple example that shows that not every sentential logic has a strongly adequate Gentzen system, a question raised in Section 2.4. Moreover, this example is interesting for other reasons.

Let us consider the $\Box$-fragment of the weak version of the normal modal logic K considered in Jansana [1991]. Let us call it just $\mathcal{S}$. The consequence relation of $\mathcal{S}$ is trivial in the sense that $\Gamma \vdash_{\mathcal{S}} \varphi$ if and only if $\varphi \in \Gamma$. It follows that $\mathcal{S}$ is non-protoalgebraic, selfextensional and non-Fregean. Moreover, any subset of any algebra is an $\mathcal{S}$-filter; since for any A the abstract logic $\langle A, P(A) \rangle$ is reduced, it follows that the class $\mathbf{Alg}\mathcal{S}$ is the class of all algebras with a single unary operation. In spite of this, not every abstract logic is a full model of $\mathcal{S}$.

In Jansana [1991] it is proved that an abstract logic $\langle A, C \rangle$ is a full model of $\mathcal{S}$ if and only if the following conditions hold:

(1) $\Box[C(X)] \subseteq C(\Box[X])$.
(2) $C(\emptyset) = \emptyset$.
(3)   For all $a, b \in A$, $a \in C(b)$ if and only if $b \in C(a)$.
(4)   If $a \in C(X)$ then there is $b \in X$ such that $a \in C(b)$.

From this it follows that $\mathcal{S}$ is strongly selfextensional.

However, condition (4) above is not directly expressible as a Gentzen-style rule, which suggests that this logic might not have a strongly adequate Gentzen system. And this is indeed the case. The reason lies in the fact that the class of full models of this logic is not closed under (finitary) direct products while the class of finitary models of any Gentzen system is always closed under this operation, as is easily checked[36].

5.4. Other miscellaneous examples

We review in this section the study of a few more sentential logics from the point of view of the determination of their $\mathcal{S}$-algebras and their full models. Three of these examples have an interesting common feature. It so happens that several of the logics mentioned in Font [1993] as examples of algebraizable logics which are not selfextensional do have a weak version which is not algebraizable but which is selfextensional; and the two logics of each pair have the same class of $\mathcal{S}$-algebras, and (of course) different classes of full models, with some characteristic

---

[36]This idea has been further developed in Font, Jansana, and Pigozzi [2006], where the following result has been obtained (Theorem 3.24): A sentential logic has a strongly adequate Gentzen system if and only if its class of full models is closed under substructures and reduced products.

relationship between them. We have seen in Section 5.3 that this is the case of the strong and the weak version of a normal modal logic, and it is easy to imagine that a parallel behaviour would be found for classical modal logics. Some further cases where this situation appears are included here; while the difference between the strong and the weak version of the logic lies often in an inference rule, the relevance logic considered below is an exception.

5.4.1. Two relevance logics

By the "system R of relevance logic" one normally understands the set of theorems of the language $(\wedge, \vee, \rightarrow, \neg)$ generated from axioms R1–R13 of Anderson and Belnap [1975] p. 341 and the rules of Modus Ponens and Adjunction $\{\varphi, \psi\} \vdash \varphi \wedge \psi$. The same axioms and rules define in the usual way a notion of consequence from premisses, that is, a sentential logic, sometimes called "official deducibility" in the literature, and also denoted by R. This logic has been shown in Blok and Pigozzi [1989a] to be algebraizable, while the R-algebras have been found in Font and Rodríguez [1990], where they are called precisely R-*algebras*; they are the De Morgan semigroups considered in p. 357 of Anderson and Belnap [1975] that satisfy $\big((a \rightarrow a) \wedge (b \rightarrow b)\big) \rightarrow c \leqslant c$ for all a, b, c; the class of De Morgan monoids, which has usually been taken as the algebraic counterpart of R at the cost of adding a truth constant $\top$ to the language, is a proper subclass of the class of R-algebras.

However, there are several reasons that suggest the consideration of a different notion of deducibility associated with the system R, that is, another sentential logic, which we will denote by WR. It is defined from the set of theorems of R as follows: For any $\Gamma \subseteq Fm$, $\varphi \in Fm$,

$$\Gamma \vdash_{\mathrm{WR}} \varphi \iff \text{There are } n > 0 \text{ and } \varphi_1, \ldots, \varphi_n \in \Gamma \text{ such that}$$
$$(\varphi_1 \wedge \cdots \wedge \varphi_n) \rightarrow \varphi \text{ is a theorem of R.}$$

Note that this implies that WR is finitary and has no theorems. This definition has been suggested by Wójcicki in Section 2.10 of [1988] as a means of obtaining a sentential logic more coherent with the idea of entailment than by simply extending the formal system for the theorems of R to deducibility from premisses; it coincides with the entailment relation associated with the ternary relational semantics of Routley, Meyer and Fine, as follows from their completeness theorems, see Anderson, Belnap, and Dunn [1992] Sections 48, 51. Indeed, WR satisfies the following version of the so-called Relevance Principle or Variable-Sharing Property: If $\varphi \vdash_{\mathrm{WR}} \psi$ then φ and ψ must share at least one propositional variable.

In Rodríguez [1990] and in Font and Rodríguez [1994] the two logics R and WR are studied from the point of view of the present monograph. It is proved that WR is non-protoalgebraic, that it is selfextensional and not Fregean, and that R is not selfextensional. Actually R is the axiomatic extension of WR determined by the *Identity Law*, $\varphi \to \varphi$, as additional axiom scheme. The WR-algebras are also the R-algebras. The full models of WR are found: They are the abstract logics whose $(\wedge, \vee, \neg)$-reduct is a full model of Belnap's logic DM and that satisfy the following four additional conditions relating the closure operator C to $\to$; the second one is the residuation property of implication with respect the binary connective $a * b = \neg(a \to \neg b)$, usually called "fusion" or "multiplicative conjunction":

(1) $b \in C(a, a \to b)$.
(2) $c \in C(a * b) \iff b \to c \in C(a)$.
(3) $b \to (a \to c) \in C\big(a \to (b \to c)\big)$.
(4) $c \in C\big(((a \to a) \wedge (b \to b)) \to c\big)$

Since all these logics have the congruence property, WR is an example of a strongly selfextensional but neither Fregean nor protoalgebraic logic. The full models of R are characterized as the axiomatic extensions of full models of WR by the Identity Law: An abstract logic $\mathbb{L} = \langle A, C \rangle$ is a full model of R iff there is a full model of WR, $\mathbb{L}_0 = \langle A, C_0 \rangle$, such that $\mathcal{C} = \{T \in \mathcal{C}_0 : \forall x \in A, \ x \to x \in T\}$. Moreover, WR is an example of a non-protoalgebraic logic with $\mathbf{Alg}^*\mathcal{S} = \mathbf{Alg}\mathcal{S}$; see the discussion on page 62.

Finally let us mention that in Font and Rodríguez [1994] a Gentzen system for WR is presented and proved to be strongly adequate for it. Since WR is selfextensional and satisfies the PC, all results of Section 4.2 apply. The presentation of this Gentzen system is the one for Belnap's logic mentioned in Section 5.1.3 augmented with two axioms corresponding to conditions (3) and (4) above, and with three rules, corresponding to conditions (1) and (2).

5.4.2. Sette's paraconsistent logic

The so-called "maximal paraconsistent logic" P^1 was introduced and first studied in Sette [1973]. Its primitive connectives are $\neg$ and $\to$. Its axioms are the following:
$$\varphi \to (\psi \to \varphi)$$
$$\big(\varphi \to (\psi \to \xi)\big) \to (\varphi \to \psi) \to (\varphi \to \xi)$$
$$(\neg\varphi \to \neg\psi) \to \big((\neg\varphi \to \neg\neg\psi) \to \varphi\big)$$
$$\neg(\varphi \to \neg\neg\varphi) \to \varphi$$
$$(\varphi \to \psi) \to \neg\neg(\varphi \to \psi)$$

Its only rule of inference is Modus Ponens. It is semantically determined by a three-valued matrix. It is a *paraconsistent logic*, i.e., in it a theory containing both φ and $\neg\varphi$ for some formula φ is not necessarily inconsistent; and it is *maximal* in the sense that its only proper non-trivial axiomatic extension is CPC. It was proved to be algebraizable in Lewin, Mikenberg, and Schwarze [1990]; the associated class of algebras, which is the quasivariety generated by the three-element algebra being the reduct of the characteristic matrix of the logic, was studied in Lewin, Mikenberg, and Schwarze [1994] and independently in Pynko [1995a]. This class is a proper quasivariety, called the class of $\mathbf{P}^1$-*algebras* in the former paper, and the class of *Sette algebras* in the latter. In this last paper the logic $\mathbf{P}^1$ is also studied from the point of view of abstract logics. There it is also proved that $\mathbf{P}^1$ is not regularly algebraizable, and that the abstract logics associated with it (its full models in our terminology) can be characterized as the finitary models of $\mathbf{P}^1$ that satisfy the DDT with respect to $\rightarrow$. It is interesting to remark the similarity of this result to our Corollary 4.48: There, from the assumption that a logic is selfextensional and has the DDT, it is proved that its full models are exactly its finitary models satisfying the DDT and the congruence property; in spite of the fact that $\mathbf{P}^1$ is not selfextensional, as we show below, we get an analogous characterization, without the congruence property, by an ad-hoc proof rather than from a general argument.

The reason why $\mathbf{P}^1$ is not selfextensional is the following: If it were so, by our Theorem 4.46 it would be strongly selfextensional, because it satisfies the DDT with respect to some connective. Then by our Proposition 3.20 it would be Fregean and protoalgebraic, and since it has theorems by definition, Theorem 3.18 implies that it would be regularly algebraizable, and Proposition 4.49 implies that it would be strongly algebraizable; but both things are shown to be false in Pynko [1995a].

Some new connectives can be introduced (we follow Pynko's definition in his [1995a], which differs from Sette's): First a new negation $\tilde{\neg}\varphi = \varphi \rightarrow \neg(\varphi \rightarrow \varphi)$, and from it as in classical logic one defines $\varphi \vee \psi = \tilde{\neg}\varphi \rightarrow \psi$ and $\varphi \wedge \psi = \tilde{\neg}(\tilde{\neg}\varphi \vee \tilde{\neg}\psi)$, and the full models of $\mathbf{P}^1$ are *classical* with respect to these connectives, that is, they satisfy the PRA with respect to $\tilde{\neg}$, the PC with respect to $\wedge$ and the PDI with respect to $\vee$. The converse is not true: Pynko has shown (in a personal communication) a four-element algebra with an abstract logic that satisfies all these properties but is not a full model of the logic $\mathbf{P}^1$; actually, it is not even a model of this logic.

5.4.3. Tetravalent modal logic

This little known sentential logic is a modal extension of Belnap's four-valued logic, and is related to the class of *tetravalent modal algebras*. These algebras were defined by Monteiro, as a weakening of *three-valued Łukasiewicz algebras*, and they have been studied mainly by Loureiro (see Loureiro [1982], [1985] among others), and by Figallo [1992] under a slightly different name. Abstract logics related to the logic and the algebras were initially studied in Font and Rius [1990] and in Rius [1992], and more specifically from the present point of view in Font and Rius [2000]. This case is especially interesting because its behaviour presents at the same time some distinctive features of Belnap's four-valued logic, such as some semantical characterizations or the Gentzen systems, and some of the normal modal logics, such as the interplay between the strong and the weak versions due to the Rule of Necessitation.

As in the last group, we find two versions of the logic: The weak one, called TML and defined by a Gentzen system, is protoalgebraic and finitely equivalential, but is not algebraizable; it is, however, selfextensional and non-Fregean. The strong one, TML_N is obtained from the weak one after the addition of the full Rule of Necessitation, and is algebraizable but not selfextensional; the defining equation is $p \approx p \dagger p$ and the equivalence formula is $p \dagger q$, where

$$\varphi \dagger \psi = \big[\neg\Box(\varphi \vee \psi) \vee \Box(\varphi \wedge \psi)\big] \wedge \big[\Box\neg(\varphi \vee \psi) \vee \neg\Box\neg(\varphi \wedge \psi)\big]$$

is a term which plays an important role both in the logic and in the algebraic theory of tetravalent modal algebras. It has the additional interest of being an example of an equivalence connective for a logic which does not seem to be, at least in an obvious way, the result of the "symmetrization" of an implication connective that plays a significant role in the logic. For both logics the class of S-algebras is the class of tetravalent modal algebras, and the full models of TML are the full models of DM satisfying additional properties concerning $\Box$, while for TML_N they are the strong versions of the former, in a sense similar to that of Section 5.3.

The variety of tetravalent modal algebras, as in the case of De Morgan algebras, is generated by a four-element algebra, and this algebra also generates the two logics by taking on it either the matrix with only the maximum in the filter, for TML_N, or the generalized matrix consisting of the two prime filters of the lattice, for TML. It was proved in Font and Rius [1990] that in this case, the full models of TML can be characterized as those abstract logics projectively generated from this generalized matrix by families of homomorphisms of a specified form. The usual theorem in the form of 2.21 was also obtained; the full models of TML are those finitary abstract logics whose reduction consists of a tetravalent modal algebra and all its filters, while in the case of TML_N one takes the open filters. Finally, since the full models of TML can be characterized by conditions on the closure operator corresponding to the Gentzen system, TML is strongly self-extensional and the Gentzen system is strongly adequate for it. Since this logic satisfies the PC, the results of Section 4.2 apply.

5.4.4. Logics related to cardinality restrictions in the Deduction Theorem

The many attempts in the literature to find more general versions of the Herbrand-Tarski Deduction Theorem have concentrated in generalizing the implication connective to a finite or arbitrary set of formulas, possibly with *parameters*, and making it *local*; see Blok and Pigozzi [1991], Czelakowski [1986], Czelakowski and Dziobiak [1991]. Here we review some work done on weakened versions of the DT (the MP is always assumed) along quite a different line, namely by making its validity depend on the cardinality of the set Γ of supplementary premisses that appears in the DT; some material concerning this topic is included in García Lapresta [1991][37] and was partly anticipated in García Lapresta [1988b], [1988a]; the first published source known to us where this kind of weakenings is considered is Pla and Verdú [1980].

The easiest way to obtain logics satisfying such limited versions of the DT is to define them through a suitable Gentzen system having the intended property as a primitive rule. Consider the Gentzen system $\mathfrak{G}_n$, of type ω in the language $(\rightarrow)$ with the structural rules of Definition 4.1 and the two rules (MP) and (DTn) as introduced in Section 5.2.1. We already know that the logic defined by $\mathfrak{G}_n$ is precisely $\text{IPC}_\rightarrow$ when $n \geqslant 2$.

The case $n = 1$ is more interesting. The primitive non-structural rules of $\mathfrak{G}_1$ are (MP) as in Section 5.2.1 and (DT1); recall that (DT1) is actually the union of the two rule schemas:

$$(\text{DT0}) \quad \frac{\varphi \vdash \psi}{\vdash \varphi \rightarrow \psi} \qquad\qquad (\text{DT1}') \quad \frac{\xi, \varphi \vdash \psi}{\xi \vdash \varphi \rightarrow \psi}$$

Call $\mathcal{G}_1$ the sentential logic defined by this Gentzen system. This logic is protoalgebraic but not algebraizable, because it is not equivalential, and it is self-extensional but not Fregean. A kind of Hilbert-style presentation of $\mathcal{G}_1$ has the following axiom schema and rules of inference:

$$(\text{K}) \quad \varphi \rightarrow (\psi \rightarrow \varphi)$$

$$(\text{MP}) \quad \{\varphi, \varphi \rightarrow \psi\} \vdash \psi$$

---

[37] Some of the facts mentioned in this section have not been published until Bou, Font, and García Lapresta [2004], along with a few others.

$$(\text{R-MP2}) \quad \frac{\vdash \eta \to (\xi \to \varphi) \qquad \vdash \eta \to (\xi \to (\varphi \to \psi))}{\vdash \eta \to (\xi \to \psi)}$$

Note that (MP) is unrestricted but (R-MP2), which in some sense is a strengthening of Modus Ponens, is restricted to theorems. Strictly speaking, this is not a Hilbert-style presentation of the consequence relation of the logic, but only of its theorems; but the theories of $\mathcal{G}_1$ are the sets of formulas containing its theorems and closed under (MP).

The algebraization of the Gentzen system $\mathfrak{G}_1$ is straightforward, because it satisfies the congruence property, and hence the equations of its reduced models are expressed directly by the closure operator. Then $\mathbf{Alg}\mathfrak{G}_1 = \mathbf{QH}$, the class of *quasi-Hilbert algebras* introduced in Pla and Verdú [1980]: These are algebras $A = \langle A, \to \rangle$ of type (2) such that there is an element $1 \in A$ satisfying, for all $a, b, c, d \in A$:

(QH1) $a \to b = b \to a = 1$ implies $a = b$;

(QH2) $a \to (b \to a) = 1$; and

(QH3) $a \to (b \to c) = a \to (b \to (c \to d)) = 1$ implies $a \to (b \to d) = 1$.

This quasivariety is larger than the variety of Hilbert algebras but smaller than the class of implicative algebras[38].

A sentential logic whose algebraization is exactly the class $\mathbf{QH}$ is the "strong version" of $\mathcal{G}_1$, that is, the logic whose only axiom is (K) and whose rules are (MP) and the unrestricted version of (R-MP2), that is, the rule

(MP2) $\{\eta \to (\xi \to \varphi), \eta \to (\xi \to (\varphi \to \psi))\} \vdash \eta \to (\xi \to \psi).$

Let us call this logic $\mathcal{H}_1$. It is an extension of $\mathcal{G}_1$; actually its theories are exactly those of $\mathcal{G}_1$ that are closed under (MP2). It follows that $\mathcal{H}_1$ and $\mathcal{G}_1$ have the same theorems, and it can be proved that $\mathcal{H}_1$ is regularly algebraizable, with the defining equation $p \approx p \to p$ and equivalence formulas $\{p \to q, q \to p\}$, that it is not selfextensional, and that it does not satisfy any of the (DTn), not even the weakest (DT0); this implies that $\mathcal{G}_1$ is weaker than $\mathcal{H}_1$. The equivalent quasivariety semantics of $\mathcal{H}_1$ is $\mathbf{QH}$, with $\{1\}$ as the filter of the corresponding reduced matrix. Since $\mathbf{QH}$ is larger than the class of Hilbert algebras, we know that $\mathcal{H}_1$, and hence $\mathcal{G}_1$, are weaker than $\text{IPC}_{\to}$.

Since $\mathcal{H}_1$ is algebraizable, by Corollary 3.11 we know that the full models of $\mathcal{H}_1$ are exactly determined by the families of all the $\mathcal{H}_1$-filters containing a given one. From the Hilbert-style definition of the logic we see that if $A \in \mathbf{QH}$ then a

---

[38] It is not known whether this quasivariety is actually a variety. If it is not, then $\mathfrak{G}_1$ would not be strongly adequate for $\mathcal{G}_1$, because in Bou, Font, and García Lapresta [2004] it is proved that $\mathbf{Alg}\mathcal{G}_1$ is the variety generated by $\mathbf{GH}$. In the same paper the full models of $\mathcal{G}_1$ are characterized.

subset $D \subseteq A$ is an $\mathcal{H}_1$-filter if and only if $1 \in D$ and D is closed under (MP) and (MP2).

The case $n = 0$ is slightly different, since (DT0) is really very weak; for instance it does not imply congruence. It is known that the logic defined by $\mathfrak{G}_0$ is protoalgebraic but not equivalential (hence it is not algebraizable) and also that it is not selfextensional. As in the case $n = 1$, a Hilbert-style presentation with restricted rules has been produced, but in contrast the corresponding strong version is not algebraizable.

The following extension of $\mathfrak{G}_0$ will yield completely parallel results to those obtained for $\mathfrak{G}_1$. The rules to be added are the rule of prefixing and a restricted rule of congruence that already appears in Rasiowa [1974] p. 213:

$$\text{(PR)} \quad \frac{\Gamma \vdash \varphi}{\Gamma \vdash \psi \to \varphi} \qquad\qquad \text{(R-C)} \quad \frac{\vdash \varphi \to \psi \qquad \vdash \varphi' \to \psi'}{\vdash (\psi \to \varphi') \to (\varphi \to \psi')}$$

To be precise, let us call $\mathfrak{G}^1$ the Gentzen system of type ω in the language $(\to)$ whose rules are (MP), (DT0), (PR) and (R-C), in addition to the structural ones; this Gentzen system is closer to $\mathfrak{G}_1$ than to $\mathfrak{G}_0$, hence the name we have given to it. Call $\mathcal{G}^1$ the sentential logic defined by this Gentzen system. This logic is protoalgebraic but not equivalential (thus, it is not algebraizable), and it is selfextensional but not Fregean. The pseudo-Hilbert-style presentation of $\mathcal{G}^1$ has the single axiom schema (K) and three rules of inference, the unrestricted rule (MP) and the other two rules restricted to theorems:

$$\text{(R-MP1)} \quad \frac{\vdash \xi \to \varphi \qquad \vdash \xi \to (\varphi \to \psi)}{\vdash \xi \to \psi}$$

and (R-C) taken as a rule on theorems, as above. It is not difficult to show that $\mathbf{Alg}\mathfrak{G}^1 = \mathbf{QH}^1$, the class of algebras $\boldsymbol{A} = \langle A, \to \rangle$ of type (2) having an element 1 satisfying the axioms (QH1) and (QH2) of quasi-Hilbert algebras and moreover, for all $a, b, c, d \in A$:

(QH4) $a \to b = a \to (b \to c) = 1$ implies $a \to c = 1$; and

(QH5) $a \to b = c \to d = 1$ implies $(b \to c) \to (a \to d) = 1$.

This quasivariety is larger than $\mathbf{QH}$, but it is still smaller than the class of implicative algebras.

A sentential logic whose algebraization is exactly the class $\mathbf{QH}^1$ is the "strong version" of $\mathcal{G}^1$, that is, the logic whose only axiom is (K) and whose rules are (MP) and the unrestricted versions of (R-C) and (R-MP1), that is, the rules

(C) $\{\varphi \to \psi, \varphi' \to \psi'\} \vdash (\psi \to \varphi') \to (\varphi \to \psi')$

(MP1) $\{\xi \to \varphi, \xi \to (\varphi \to \psi)\} \vdash \xi \to \psi.$

Call $\mathcal{H}^1$ this logic. Clearly it is an extension of $\mathcal{G}^1$, since its theories are those of $\mathcal{G}^1$ that are closed under MP1 and C. It follows that $\mathcal{H}^1$ and $\mathcal{G}^1$ have the same theorems, and it can be proved that $\mathcal{H}^1$ is regularly algebraizable (with the same defining equation and equivalence formulas as $\mathcal{H}_1$) but not selfextensional, and that it does not satisfy any of the DTn. Its equivalent quasivariety semantics is $\mathbf{QH^1}$, and $\{1\}$ is the least $\mathcal{H}^1$-filter on any algebra in this class. Using ad-hoc matrices and the fact that $\mathbf{QH^1}$ is larger than $\mathbf{QH}$ one can prove that $\mathcal{G}^1$ is weaker than $\mathcal{H}^1$ and also than $\mathcal{G}_1$, and that $\mathcal{H}^1$ is weaker than $\mathcal{H}_1$.

Since $\mathcal{H}^1$ is algebraizable, by Corollary 3.11 we know that the full models of $\mathcal{H}^1$ are exactly determined by the families of all the $\mathcal{H}^1$-filters containing a given one. From the Hilbert-style definition of the logic we see that if $A \in \mathbf{QH^1}$ then a subset $D \subseteq A$ is an $\mathcal{H}^1$-filter if and only if $1 \in D$ and D is closed under the rules (MP), (C) and (MP1).

BIBLIOGRAPHY

R. ADILLON AND V. VERDÚ

[1996] *A Gentzen system equivalent to the BCK-logic*, **Bulletin of the Section of Logic**, vol. 25, no. 2, pp. 73–79.

A. R. ANDERSON AND N. D. BELNAP

[1975] *Entailment. The logic of relevance and necessity*, vol. I, Princeton University Press.

A. R. ANDERSON, N. D. BELNAP, AND J. M. DUNN

[1992] *Entailment. The logic of relevance and necessity*, vol. II, Princeton University Press.

S. BABYONYSHEV

[2003] *Fully Fregean logics*, **Reports on Mathematical Logic**, vol. 37, pp. 59–78.

R. BALBES AND P. DWINGER

[1974] *Distributive lattices*, University of Missouri Press, Columbia (Missouri).

F. BELARDINELLI, P. JIPSEN, AND H. ONO

[2004] *Algebraic aspects of cut elimination*, **Studia Logica**, vol. 77, pp. 209–240.

N. D. BELNAP

[1977] *A useful four-valued logic*, **Modern uses of multiple-valued logic** (J. M. Dunn and G. Epstein, editors), Reidel, Dordrecht-Boston, pp. 8–37.

G. BIRKHOFF

[1973] *Lattice theory*, 3rd. ed., Colloquium Publications, vol. XXV, American Mathematical Society, Providence, (1st. ed. 1940).

W. BLOK AND B. JÓNSSON

[2006] *Equivalence of consequence operations*, **Studia Logica (Special issue in memory of Willem Blok)**, vol. 83, pp. 91–110.

W. J. BLOK AND P. KÖHLER

[1983] *Algebraic semantics for quasi-classical modal logics*, **The Journal of Symbolic Logic**, vol. 48, pp. 941–964.

W. J. BLOK AND D. PIGOZZI

[1986] *Protoalgebraic logics*, **Studia Logica**, vol. 45, pp. 337–369.

[1988] *Alfred Tarski's work on general metamathematics*, **The Journal of Symbolic Logic**, vol. 53, pp. 36–50.

[1989a] **Algebraizable logics**, Memoirs of the American Mathematical Society, vol. 396, A.M.S., Providence, January.

[1989b] The Deduction Theorem in algebraic logic, Unpublished manuscript, partly subsumed in Blok and Pigozzi [200x].

[1991] *Local deduction theorems in algebraic logic*, **Algebraic logic** (H. Andréka, J. D. Monk, and I. Németi, editors), Colloquia Mathematica Societatis János Bolyai, vol. 54, North-Holland, Amsterdam, pp. 75–109.

[1992] *Algebraic semantics for universal Horn logic without equality*, **Universal algebra and quasigroup theory** (A. Romanowska and J. D. H. Smith, editors), Heldermann, Berlin, pp. 1–56.

[200x] *Abstract algebraic logic and the Deduction Theorem*, **The Bulletin of Symbolic Logic**, To appear.

S. L. BLOOM

[1977] *A note on Ψ-consequences*, **Reports on Mathematical Logic**, vol. 8, pp. 3–9.

S. L. BLOOM AND D. J. BROWN

[1973] *Classical abstract logics*, **Dissertationes Mathematicae (Rozprawy Mat.)**, vol. 102, pp. 43–51.

F. BOU

[2001] *Implicación estricta y lógicas subintuicionistas*, **Master Thesis**, University of Barcelona.

F. BOU, F. ESTEVA, J. M. FONT, A. GIL, LL. GODO, A. TORRENS, AND V. VERDÚ

[2009] *Logics preserving degrees of truth from varieties of residuated lattices*, **Journal of Logic and Computation**, To appear.

F. BOU, J. M. FONT, AND J. L. GARCÍA LAPRESTA

[2004] *On weakening the deduction theorem and strengthening modus ponens*, **Mathematical Logic Quarterly**, vol. 50, pp. 303–324.

D. J. BROWN

[1969] *Abstract logics, **Ph. D. Thesis***, Stevens Institute of Technology.

D. J. BROWN AND R. SUSZKO

[1973] *Abstract logics*, **Dissertationes Mathematicae (Rozprawy Mat.)**, vol. 102, pp. 9–42.

S. BURRIS AND H. P. SANKAPPANAVAR

[1981] *A course in universal algebra*, Springer-Verlag, New York.

E. CASANOVAS, P. DELLUNDE, AND R. JANSANA

[1996] *On elementary equivalence for equality-free logic*, **Notre Dame Journal of Formal Logic**, vol. 37, no. 3, pp. 506–522.

S. CELANI AND R. JANSANA

[2001] *A closer look at some subintuitionistic logics*, **Notre Dame Journal of Formal Logic**, vol. 42, pp. 225–255.

B. F. CHELLAS

[1980] **Modal logic: An introduction**, Cambridge University Press, Cambridge, Cambridge.

R. CIGNOLI

[1991] *Quantifiers on distributive lattices*, **Discrete Mathematics**, vol. 96, pp. 183–197.

J. CZELAKOWSKI

[1980] *Reduced products of logical matrices*, **Studia Logica**, vol. 39, pp. 19–43.

[1981] *Equivalential logics, I, II*, **Studia Logica**, vol. 40, pp. 227–236 and 355–372.

[1984] *Remarks on finitely based logics*, **Models and Sets** (G. H. Müller and M. M. Richter, editors), Lecture Notes in Mathematics, vol. 1103, Springer Verlag, Berlin, pp. 147–168.

[1985] *Algebraic aspects of deduction theorems*, **Studia Logica**, vol. 44, pp. 369–387.

[1986] *Local deductions theorems*, **Studia Logica**, vol. 45, pp. 377–391.

[1992] *Consequence operations: Foundational studies*, **Reports of the Research Project "Theories, Models, Cognitive Schemata"**, Institute of Philosophy and Sociology, Polish Academy of Sciences, Warszawa.

[2001a] *Protoalgebraic logics*, Trends in Logic, Studia Logica Library, vol. 10, Kluwer Academic Publishers, Dordrecht.

[2001b] *Protoalgebraic logics*, Trends in Logic - Studia Logica Library, vol. 10, Kluwer Academic Publishers, Dordrecht.

J. CZELAKOWSKI AND W. DZIOBIAK

[1991] *A deduction theorem schema for deductive systems of propositional logics*, **Studia Logica, Special Issue on Algebraic Logic**, vol. 50, pp. 385–390.

J. CZELAKOWSKI AND R. JANSANA

[2000] *Weakly algebraizable logics*, **The Journal of Symbolic Logic**, vol. 65, no. 2, pp. 641–668.

J. CZELAKOWSKI AND G. MALINOWSKI

[1985] *Key notions of Tarski's methodology of deductive systems*, **Studia Logica**, vol. 44, pp. 321–351.

J. CZELAKOWSKI AND D. PIGOZZI

[2004a] *Fregean logics*, **Annals of Pure and Applied Logic**, vol. 127, pp. 17–76.

[2004b] *Fregean logics with the multiterm deduction theorem and their algebraization*, **Studia Logica**, vol. 78, pp. 171–212.

P. DELLUNDE

[1996] *Contributions to the model theory of equality-free logic*, **Ph. D. Thesis**, University of Barcelona.

[1999] *Equality-free logic: The method of diagrams and preservation theorems*, **Logic Journal of the IGPL**, vol. 7, pp. 717–732.

[2000a] *On definability of the equality in classes of algebras with an equivalence relation*, **Studia Logica**, vol. 64, pp. 345–353.

[2000b] *A preservation theorem for equality-free Horn sentences*, **Theoria (San Sebastián)**, vol. 39, pp. 517–530.

[2003] *Equality-free saturated models*, **Reports on Mathematical Logic**, vol. 37, pp. 3–22.

P. DELLUNDE AND R. JANSANA

[1994] *On structural equivalence of theories*, Manuscript.

[1996] *Some characterization theorems for infinitary universal Horn logic without equality*, **The Journal of Symbolic Logic**, vol. 61, no. 4, pp. 1242–1260.

A. DIEGO

[1965] *Sobre álgebras de Hilbert*, Notas de Lógica Matemática, vol. 12, Universidad Nacional del Sur, Bahía Blanca (Argentina).

[1966] *Sur les algèbres de Hilbert*, Gauthier-Villars, Paris.

K. DOŠEN

[1986] *Modal translations and intuitionistic double negation*, **Logique et Analyse**, vol. 29, pp. 81–94.

K. DOŠEN AND P. SCHROEDER-HEISTER

[1993] *Substructural logics*, Studies in Logic and Computation, vol. 2, Oxford University Press.

J. M. DUNN

[1976] *Intuitive semantics for first-degree entailments and 'coupled trees'*, **Philosophical Studies**, vol. 29, pp. 149–168.

J. M. DUNN AND G. RESTALL

[2002] *Relevance logic and entailment*, **Handbook of philosophical logic, second edition** (D. Gabbay and F. Guenthner, editors), vol. 6, Kluwer, Dordrecht, pp. 1–128.

K. DYRDA AND T. PRUCNAL

[1980] *On finitely based consequence determined by a distributive lattice*, **Bulletin of the Section of Logic**, vol. 9, pp. 60–66.

R. ELGUETA

[1994] *Algebraic model theory for languages without equality*, **Ph. D. Thesis**, University of Barcelona.

[1997] *Characterizing classes defined without equality*, **Studia Logica**, vol. 58, no. 3, pp. 357–394.

[1998] *Subdirect representation theory for classes without equality*, **Algebra Universalis**, vol. 40, pp. 201–246.

R. ELGUETA AND R. JANSANA

[1999] *Definability of Leibniz equality*, **Studia Logica**, vol. 63, pp. 223–243.

A. FIGALLO

[1992] *On the congruences in four-valued modal algebras*, **Portugaliae Mathematica**, vol. 49, pp. 249–261.

J. M. FONT

[1980] *Introducció d'interiors d'ordre en lògiques abstractes*, **Publicacions de la Secció de Matemàtiques, Universitat Autònoma de Barcelona**, vol. 20, pp. 79–82.

[1987] *On some congruence lattices of a topological Heyting lattice*, **Contributions to general algebra** (J. Czermak et al., editors), vol. 5, Teubner, Stuttgart, pp. 129–137.

[1993] *On the Leibniz congruences*, **Algebraic methods in logic and in computer science** (C. Rauszer, editor), Banach Center Publications, vol. 28, Polish Academy of Sciences, Warszawa, pp. 17–36.

[1997] *Belnap's four-valued logic and De Morgan lattices*, **Logic Journal of the I.G.P.L.**, vol. 5, no. 3, pp. 413–440.

[2003a] *An abstract algebraic logic view of some multiple-valued logics*, **Beyond two: Theory and applications of multiple-valued logic** (M. Fitting and E. Orlowska, editors), Studies in Fuzziness and Soft Computing, vol. 114, Physica-Verlag, Heidelberg, pp. 25–58.

[2003b] *Generalized matrices in abstract algebraic logic*, **Trends in logic. 50 years of studia logica** (V. F. Hendriks and J. Malinowski, editors), Trends in Logic - Studia Logica Library, vol. 21, Kluwer, Dordrecht, pp. 57–86.

[2006] *Beyond Rasiowa's algebraic approach to non-classical logics*, **Studia Logica**, vol. 82, pp. 172–209.

J. M. FONT, A. GIL, A. TORRENS, AND V. VERDÚ

[2006] *On the infinite-valued Łukasiewicz logic that preserves degrees of truth*, **Archive for Mathematical Logic**, vol. 45, pp. 839–868.

J. M. FONT, F. GUZMÁN, AND V. VERDÚ

[1991] *Characterization of the reduced matrices for the $\{\wedge, \vee\}$-fragment of classical logic*, **Bulletin of the Section of Logic**, vol. 20, pp. 124–128.

J. M. FONT AND R. JANSANA

[1994] *On the sentential logics associated with strongly nice and semi-nice general logics*, **Bulletin of the I.G.P.L.**, vol. 2, pp. 55–76.

[1995] *Full models for sentential logics*, **Bulletin of the Section of Logic**, vol. 24, no. 3, pp. 123–131.

[2001] *Leibniz filters and the strong version of a protoalgebraic logic*, **Archive for Mathematical Logic**, vol. 40, pp. 437–465.

J. M. FONT, R. JANSANA, AND D. PIGOZZI

[2001] *Fully adequate Gentzen systems and the deduction theorem*, **Reports on Mathematical Logic**, vol. 35, pp. 115–165.

[2003] *A survey of abstract algebraic logic*, **Studia Logica (Special issue on Abstract Algebraic Logic, Part II)**, vol. 74, pp. 13–97.

[2006] *On the closure properties of the class of full g-models of a deductive system*, **Studia Logica (Special issue in memory of Willem Blok)**, vol. 83, pp. 215–278.

J. M. FONT AND M. RIUS

[1990] *A four-valued modal logic arising from Monteiro's last algebras*, **Proceedings of the 20th international symposium on multiple-valued logic** (Charlotte), The IEEE Computer Society Press, pp. 85–92.

[2000] *An abstract algebraic logic approach to tetravalent modal logics*, **The Journal of Symbolic Logic**, vol. 65, no. 2, pp. 481–518.

J. M. FONT AND G. RODRÍGUEZ

[1990] *Note on algebraic models for relevance logic*, **Zeitschrift für Mathematische Logik und Grundlagen der Mathematik**, vol. 36, pp. 535–540.

[1994] *Algebraic study of two deductive systems of relevance logic*, **Notre Dame Journal of Formal Logic**, vol. 35, no. 3, pp. 369–397.

J. M. FONT AND V. VERDÚ

[1979] *Lògiques abstractes, operadors interior, i lògiques modals S4*, **Revista de la Universidad de Santander**, vol. II, no. 2, pp. 867–869 and 1003–1015.

[1988] *Abstract characterization of a four-valued logic*, **Proceedings of the 18th international symposium on multiple-valued logic** (Palma de Mallorca), The IEEE Computer Society Press, pp. 389–396.

[1989a] *Completeness theorems for a four-valued logic related to De Morgan lattices*, **Faculty of Mathematics Preprint Series**, University of Barcelona, March, 10pp.

[1989b] *A first approach to abstract modal logics*, **The Journal of Symbolic Logic**, vol. 54, pp. 1042–1062.

[1990] *Two levels of modality: an algebraic approach*, **Logic counts** (E. Żarnecka-Biały, editor), Reidel, Dordrecht, pp. 53–62.

[1991] *Algebraic logic for classical conjunction and disjunction*, **Studia Logica, Special Issue on Algebraic Logic**, vol. 50, pp. 391–419.

D. GABBAY

[1994] **What is a logical system ?**, Studies in Logic and Computation, vol. 4, Oxford University Press.

N. GALATOS, P. JIPSEN, T. KOWALSKI, AND H. ONO

[2007] **Residuated lattices: an algebraic glimpse at substructural logics**, Studies in Logic and the Foundations of Mathematics, vol. 151, Elsevier, Amsterdam.

J. L. GARCÍA LAPRESTA

[1988a] *El Principio de la Deducción con dos o menos premisas: Estudio algebraico*, **Actas de las XIII Jornadas Hispano-Lusas de Matemáticas** (Valladolid), To appear.

[1988b] *Restricciones en la Propiedad de la Deducción: Análisis preliminar*, **Actes del VII Congrés Català de Lògica**, Barcelona, pp. 47–50.

[1991] *Lógicas finitamente deductivas. Restricciones de cardinalidad en la Propiedad de la Deducción*, **Ph. D. Thesis**, University of Barcelona.

A. J. GIL

[1996] *Sistemes de Gentzen multidimensionals i lògiques finitament valorades. Teoria i Aplicacions*, **Ph. D. Thesis**, University of Barcelona.

A. J. GIL, A. TORRENS, AND V. VERDÚ

[1997] *On Gentzen systems associated with the finite linear MV-algebras*, **Journal of Logic and Computation**, vol. 7, no. 4, pp. 473–500.

A. GRZEGORCZYK

[1972] *An approach to logical calculi*, **Studia Logica**, vol. 30, pp. 33–43.

R. HARROP

[1965] *Some structure results for propositional calculi*, **The Journal of Symbolic Logic**, vol. 30, pp. 271–292.

[1968] *Some forms of models of propositional calculi*, **Contributions to mathematical logic** (H. A. Schmidt, K. Schütte, and H. J. Thiele, editors), North-Holland, pp. 163–174.

B. HERRMANN

[1993a] *Algebraizability and Beth's Theorem for equivalential logics*, **Bulletin of the Section of Logic**, vol. 22, pp. 85–88.

[1993b] *Equivalential logics and definability of truth*, **Ph. D. Thesis**, Freie Universität Berlin, 61 pp.

B. HERRMANN AND F. WOLTER

[1994] *Representations of algebraic lattices*, **Algebra Universalis**, vol. 31, pp. 612–613.

R. JANSANA

[1991] *Los fragmentos □ de la lógica modal K*, **Actas del VII Congreso de Lenguajes Naturales y Lenguajes Formales** (Vic, Barcelona) (C. Martín-Vide, editor), pp. 409–413.

[1992] *La lógica de la demostrabilidad y una semántica de lógicas abstractas*, Manuscript.

[1995] *Abstract modal logics*, **Studia Logica**, vol. 55, no. 2, pp. 273–299.

[2002] *Full models for positive modal logic*, **Mathematical Logic Quarterly**, vol. 48, pp. 427–445.

[2003] *Leibniz filters revisited*, **Studia Logica**, vol. 75, pp. 305–317.

[2005] *Selfextensional logics with implication*, **Logica universalis** (J.-Y. Béziau, editor), Birkhäuser Verlag, Basel, pp. 65–88.

[2006] *Selfextensional logics with a conjunction*, **Studia Logica**, vol. 84, pp. 63–104.

H. J. KEISLER AND A. MILLER
[2001] *Categoricity without equality*, **Fundamenta Mathematicae**, vol. 170, pp. 87–106.

D. E. KNUTH, T. LARRABEE, AND P. M. ROBERTS
[1989] **Mathematical writing**, MAA Notes Series, vol. 14, The Mathematical Association of America.

M. KRACHT
[2007] *Modal consequence relations*, **Handbook of modal logic** (P. Blackburn, J. van Benthem, and F. Wolter, editors), Studies in Logic and Practical Reasoning, vol. 3, Elsevier, Amsterdam, pp. 491–545.

E J. LEMMON
[1966] *Algebraic semantics for modal logics (I and II)*, **The Journal of Symbolic Logic**, vol. 31, pp. 46–65 and 191–218.

R. A. LEWIN, I. F. MIKENBERG, AND M. G. SCHWARZE
[1990] *Algebraization of paraconsistent logic* P^1, **Journal of Non-Classical Logic**, vol. 7, pp. 79–88.

[1991] C_1 *is not algebraizable*, **Notre Dame Journal of Formal Logic**, vol. 32, pp. 609–611.

[1994] $P1$ *algebras*, **Studia Logica**, vol. 53, pp. 21–28.

J. ŁOŚ
[1949] **O matrycach logicznych**, Ser. B, vol. 19, Prace Wrocławskiego Towarzystwa Naukowege.

J. ŁOŚ AND R. SUSZKO
[1958] *Remarks on sentential logics*, **Indagationes Mathematicae**, vol. 20, pp. 177–183.

I. LOUREIRO
[1982] *Axiomatisation et propriétés des algèbres modales tétravalentes*, **Comptes Rendus de l'Académie des Sciences de Paris, Série I, Mathématique**, vol. 295, pp. 555–557.

[1985] *Principal congruences of tetravalent modal algebras*, **Notre Dame Journal of Formal Logic**, vol. 26, pp. 76–80.

D. MAKINSON

[1977] *Review 54#65*, **Mathematical Reviews**, vol. 54.

D. PIGOZZI

[1991] *Fregean algebraic logic*, **Algebraic logic** (H. Andréka, J. D. Monk, and I. Németi, editors), Colloquia Mathematica Societatis János Bolyai, vol. 54, North-Holland, Amsterdam, pp. 473–502.

J. PLA AND V. VERDÚ

[1980] *Àlgebres quasi-Hilbertianes*, **Publicacions de la Secció de Matemàtiques, Universitat Autònoma de Barcelona**, vol. 20, pp. 97–99.

W. A. POGORZELSKI AND J. SŁUPECKI

[1960a] *Basic properties of deductive systems based on nonclassical logics, I*, **Studia Logica**, vol. 9, pp. 163–176.

[1960b] *Basic properties of deductive systems based on nonclassical logics, II*, **Studia Logica**, vol. 10, pp. 77–95.

W. A. POGORZELSKI AND P. WOJTYLAK

[1982] **Elements of the theory of completeness in propositional logic**, The Silesian University, Katowice.

M. PORĘBSKA AND A. WROŃSKI

[1975] *A characterization of fragments of the intuitionistic propositional logic*, **Reports on Mathematical Logic**, vol. 4, pp. 39–42.

G. PRIEST

[1979] *The logic of paradox*, **Journal of Philosophical Logic**, vol. 8, pp. 219–241.

A. PYNKO

[1995] *On Priest's logic of paradox*, **Journal of Applied Non-Classical Logics**, vol. 5, pp. 219–225.

[1999] *Definitional equivalence and algebraizability of generalized logical systems*, **Annals of Pure and Applied Logic**, vol. 98, pp. 1–68.

A. P. PYNKO

[1995a] *Algebraic study of Sette's maximal paraconsistent logic*, **Studia Logica**, vol. 54, pp. 89–128.

[1995b] *Characterizing Belnap's logic via De Morgan's laws*, **Mathematical Logic Quarterly**, vol. 41, no. 4, pp. 442–454.

J. RAFTERY

[2006] *Correspondences between Gentzen and Hilbert systems*, **The Journal of Symbolic Logic**, vol. 71, pp. 903–957.

H. RASIOWA

[1974] *An algebraic approach to non-classical logics*, Studies in Logic and the Foundations of Mathematics, vol. 78, North-Holland, Amsterdam.

H. RASIOWA AND R. SIKORSKI

[1953] *Algebraic treatment of the notion of satisfability*, **Fundamenta Mathematicae**, vol. 40, pp. 62–95.

W. RAUTENBERG

[1981] *2-element matrices*, **Studia Logica**, vol. 40, pp. 315–353.

[1991] *Axiomatizing logics closely related to varieties*, **Studia Logica, Special Issue on Algebraic Logic**, vol. 50, pp. 607–622.

[1993] *On reduced matrices*, **Studia Logica**, vol. 52, pp. 63–72.

J. REBAGLIATO AND V. VERDÚ

[1993] *On the algebraization of some Gentzen systems*, **Fundamenta Informaticae, Special Issue on Algebraic Logic and its Applications**, vol. 18, pp. 319–338.

[1995] *Algebraizable Gentzen systems and the Deduction Theorem for Gentzen systems*, **Mathematics Preprint Series 175**, University of Barcelona, June.

M. RIUS

[1992] *Lògiques modals tetravalents*, **Ph. D. Thesis**, University of Barcelona.

A. J. RODRÍGUEZ, A. TORRENS, AND V. VERDÚ

[1990] *Łukasiewicz logic and Wajsberg algebras*, **Bulletin of the Section of Logic**, vol. 19, pp. 51–55.

G. RODRÍGUEZ

[1990] *Àlgebres i lògiques abstractes associades al càlcul R de la rellevància*, **Ph. D. Thesis**, University of Barcelona.

A. M. SETTE

[1973] *On the propositional calculus P^1*, **Mathematica Japonica**, vol. 16, pp. 173–180.

D. J. SHOESMITH AND T. J. SMILEY

[1978] *Multiple-conclusion logic*, Cambridge University Press, Cambridge.

T. J. SMILEY

[1962] *The independence of connectives*, **The Journal of Symbolic Logic**, vol. 27, pp. 426–436.

A. TARSKI

[1930] *Über einige fundamentale Begriffe der Metamathematik*, **Comptes Rendus des Séances de la Société des Sciences et des Lettres de Varsovie, Cl. III**, vol. 23, pp. 22–29.

V. VERDÚ

[1978] *Contribució a l'estudi de certs tipus de lògiques abstractes*, **Ph. D. Thesis**, University of Barcelona.

[1979] *Lògiques distributives i Booleanes*, **Stochastica**, vol. 3, pp. 97–108.

[1985] *Some algebraic structures determined by closure operators*, **Zeitschrift für Mathematische Logik und Grundlagen der Mathematik**, vol. 31, pp. 275–278.

[1986] *On some relations between closure operators and congruences*, Manuscript.

[1987] *Logics projectively generated from* $[M] = (F_4, [\{1\}])$ *by a set of homomorphisms*, **Zeitschrift für Mathematische Logik und Grundlagen der Mathematik**, vol. 33, pp. 235–241.

R. WÓJCICKI

[1969] *Logical matrices strongly adequate for structural sentential calculi*, **Bulletin de l'Académie Polonaise des Sciences, Classe III**, vol. XVII, pp. 333–335.

[1970] *Some remarks on the consequence operation in sentential logics*, **Fundamenta Mathematicae**, vol. 68, pp. 269–279.

[1973] *Matrix approach in the methodology of sentential calculi*, **Studia Logica**, vol. 32, pp. 7–37.

[1984] **Lectures on propositional calculi**, Ossolineum, Wroclaw.

[1988] **Theory of logical calculi. Basic theory of consequence operations**, Synthese Library, vol. 199, Reidel, Dordrecht.

J. J. ZEMAN

[1973] **Modal logic. The Lewis-modal systems**, Oxford University Press.

SYMBOL INDEX

$\perp$, 56, 112

$\top$, 112

$\Box$, 118

$(\wedge \vdash)$, 109

$(\vee \vdash)$, 109

$(\vdash \wedge)$, 109

$(\vdash \vee)$, 109

$\vdash$, 76

$\vdash_{\mathcal{S}}$, 25

$\vdash_{\mathfrak{G}}$, 77

$\dashv\vdash_{\mathcal{S}}$, 26

$\mathrel{\vert\!\sim}_{\mathfrak{G}}$, 76

$\mathrel{\prec\!\vert\!\sim}_{\mathfrak{G}}$, 77

$\models_{\mathbf{K}}$, 60, 83

$\models_{\mathbf{L}}$, 31

$\models_{\mathbb{L}}$, 31

$\dashv\!\models_{\mathbf{K}}$, 83

$\mathbf{2}$, 107

δ, 76

η, 15

π, 23

π_{θ}, 21

σ, 76

φ, 15

$\varphi(p, q, r, \dots)$, 16

$\varphi(\vec{q})$, 16

$\varphi^{\boldsymbol{A}}(\vec{a})$, 16

$\varphi \approx \psi$, 16

ξ, 15

ψ, 15

ω, 76

ω°, 76

Γ, 16

$\Gamma^{\boldsymbol{A}}(\vec{a})$, 16

$\Gamma \vdash \varphi$, 76

$\Gamma \vdash_{\mathcal{S}} \varphi$, 25

$\Gamma \vdash_{\mathcal{S}} \Delta$, 26

Δ, 16

$\boldsymbol{\Delta}$, 76

$\boldsymbol{\Lambda}$, 46

$\boldsymbol{\Lambda}_{\mathrm{C}}$, 46

$\boldsymbol{\Lambda}_{\mathbb{L}}$, 46

$\boldsymbol{\Lambda}(\mathbb{L})$, 46

$\boldsymbol{\Sigma}$, 76

$\boldsymbol{\Sigma}_{\mathrm{C}}$, 79

$\boldsymbol{\Sigma} \mathrel{\vert\!\sim}_{\mathfrak{G}} \boldsymbol{\Delta}$, 77

$\boldsymbol{\Omega_A}$, 16

$\boldsymbol{\Omega_A}(F)$, 16

$\widetilde{\boldsymbol{\Omega}}$, 18

$\widetilde{\boldsymbol{\Omega}}_{\boldsymbol{A}}$, 18

$\widetilde{\boldsymbol{\Omega}}_{\boldsymbol{A}}(\mathcal{C})$, 19

$\widetilde{\boldsymbol{\Omega}}_{\boldsymbol{A}}(\mathrm{C})$, 19

$\vec{a}$, 16

a^{*}, 23

A, 15

A^*, 23

$\boldsymbol{A}$, 15

$A_{\mathbb{L}}$, 17

$\boldsymbol{A}^*$, 23

$\boldsymbol{A}^-$, 119

$\boldsymbol{A}^+$, 119

$\boldsymbol{A}/\theta$, 16

Alg$\mathfrak{G}$, 80

Alg$\mathcal{S}$, 36

Alg$^*\mathcal{S}$, 28

BCI, 105

BCK, 105

$\mathcal{C}$, 17

$\mathcal{C}_{\mathbb{L}}$, 17

$\mathcal{C}^*$, 23

$\mathcal{C}^+$, 119

$\mathcal{C}^T$, 18

$\mathcal{C}/\theta$, 21

C (closure operator), 17

C (Hilbert-style rule), 128

C_θ, 40

$C_{\boldsymbol{\Sigma}}$, 79

C_{fin}, 80

$C_{\mathbb{L}}$, 17

C_n, 105

C^*, 23

C^+, 119

C^T, 18

C/θ, 21

$C(a)$, 17

$C(X, a)$, 17

$C \leqslant C'$, 18

$\mathrm{Cn}_{\mathcal{S}}$, 26

$\mathrm{Con}\,\boldsymbol{A}$, 15

$\mathrm{Con}_{\boldsymbol{K}}\boldsymbol{A}$, 15

$\mathrm{Con}\,\mathbb{L}$, 18

$\mathrm{Con}\,\mathcal{M}$, 16

CPC, 116

$\mathrm{CPC}_{\wedge\vee}$, 39, 70, 78, 107

$\mathrm{CPC}_{\to}$, 115

D, 107

DDT, 51

DM, 111

DT (Gentzen-style rule), 96, 98

DT (property), 51

DT0, 126

DT1, 126

DT1′, 126

DTn, 117

E, 105

$\mathrm{E}_{\to}$, 105

$\mathrm{Eq}(\boldsymbol{Fm})$, 16

(Eq1), 84

(Eq2), 84

(Eq3), 84

(Eq4), 84

F/θ, 16

$\mathrm{Fi}^{\boldsymbol{A}}_{\mathcal{S}}$, 27

$\mathcal{F}i_{\mathcal{S}}\boldsymbol{A}$, 27

$\mathcal{F}i^{\star}_{\mathcal{S}}\boldsymbol{A}$, 63

Fm, 15

$\boldsymbol{Fm}$, 15

$\boldsymbol{Fm}_{\kappa}$, 24

FMod$\mathcal{S}$, 33

FMod$^*\mathcal{S}$, 33

$\mathcal{F}\mathcal{M}od_{\mathcal{S}}\boldsymbol{A}$, 33

$\mathfrak{G}$, 76

$\mathfrak{G}_0$, 128

$\mathfrak{G}_1$, 126

$\mathfrak{G}_D$, 109

$\mathfrak{G}_L$, 110

$\mathfrak{G}_n$, 117, 126

$\mathfrak{G}_P$, 113
$\mathfrak{G}_\mathcal{S}$, 89
$\mathfrak{G}'_\mathcal{S}$, 98
$\mathfrak{G}^1$, 128
$\mathcal{G}_1$, 126
$\mathcal{G}_3$, 109
$\mathcal{G}_L$, 110
$\mathcal{G}^1$, 128
GL, 120

$\mathcal{H}_1$, 127
$\mathcal{H}^1$, 129
$\widetilde{\mathbf{H}}_A$, 40
$\mathrm{Hom}(\boldsymbol{A}, \boldsymbol{B})$, 15

Id, 17
IPC, 116
IPC$_\rightarrow$, 115, 116
IPC*, 113
IPC$^+$, 115

K (implicative axiom), 126
K (modal logic), 118
K_3, 112
K4, 119, 120
K4B, 120
KT, 118
KT4, 119
K, 15
$\mathbf{K}_\mathcal{S}$, 29
K3, 113

LJ, 39, 62
L, 23
L*, 23
Lat, 110
$\mathbb{L}$, 17
$\mathbb{L}_{\mathrm{fin}}$, 80
$\mathbb{L}$*, 23
$\mathbb{L}^-$, 119

$\mathbb{L}^+$, 119
$\mathbb{L}^S$, 119
$\mathbb{L}^T$, 18
$\mathbb{L}/\theta$, 21
$\mathbb{L} \cong \mathbb{L}'$, 21
$\mathbb{L} \leqslant \mathbb{L}'$, 18

$\boldsymbol{M}_4$, 111
$\mathcal{M}$, 16
$\mathcal{M}^*$, 17
$\mathcal{M}/\theta$, 16
M, 17
M*, 17
Matr$\mathcal{S}$, 27
Matr*$\mathcal{S}$, 27
Mod$\mathcal{S}$, 32
MP (Gentzen-style rule), 96, 117
MP (Hilbert-style rule), 127
MP (property), 51
MP1, 128
MP2, 127

p, 15
P, 17
P_ω°, 83
P^1, 123
P^1-algebras, 124
PC, 50
PCDL, 113
PDI, 54
PIRA, 55
PR, 128
PRA, 55

q, 15
$\vec{q}$, 16
QH, 127
QH^1, 128

R, 122

$R_\rightarrow$, 105
R-C, 128
R-MP1, 128
R-MP2, 127

s, 84
sq, 85
$\mathcal{S}$, 25
$\mathcal{S}_\emptyset$, 62
$\mathcal{S}_N$, 118
$\mathcal{S}^*$, 29
S1, 105
S2, 105
S3, 105
S4, 118
S5, 105, 118
$S5_\rightarrow$, 105
$\mathrm{Seq}(\boldsymbol{Fm})$, 76
$\mathrm{Seq}^\circ(\boldsymbol{Fm})$, 76
$\mathrm{Seq}(\text{☺})$, 77

t, 84
$t_\wedge$, 87
$t_\rightarrow$, 97
$\mathcal{T}h\mathcal{S}$, 26
TML, 125
TML_N, 125

Var, 15

WPDI, 110
WR, 62, 122

GENERAL INDEX

□-fragment of modal logic **K**, 121

1-equivalential logic, 71

abstract logic, 17
  reduced, 23
adequate, 78
  fully, 81
  strongly, 81
Adjunction, rule of, 122
algebra, 15
  of formulas, 15
algebraizable Gentzen system, 84
algebraizable logic, 60
  regularly, 71
  strongly, 93
  weakly, 66
almost inconsistent logic, 60
Andréka, H., 66
Axiom, 76
axiom scheme, 26
axiomatic extension, 18
axiomatic extensions of **K**, 118

Barcelona, 4, 6, 14
basic full model, 34
BCK logic, 105
Belnap, N.D.(jr.), 111
Belnap's four-valued logic, 111, 125

bilogical morphism, 20
Blok, W.J., 1, 2, 5, 10, 14, 16, 25,
  29, 59, 60, 66, 75, 83, 84, 94
Boolean algebra, 116
  finite, 108
  monadic, 118
  topological, 118
Brown, D.J., 5, 20

category, 44, 49
classical logic, 62
  equivalential fragment, 106
  implicative fragment, 115
classical propositional logic, 116
Classical Reductio ad Absurdum, 55
closed set
  of a Gentzen system, 79
  of an abstract logic, 17
closure algebra, 118
closure operator, 17
  finitary, 17
  structural, 26
closure system, 17
  inductive, 17
compatibility property, 59
compatible congruence, 16
complete, 30, 32
Completeness Theorem, 30, 38, 79

congruence, 15
  compatible, 16
  logical, 18
  of a matrix, 16
Congruence (rule of $\models_K$), 83
congruence property, 47
congruence rules (of a Gentzen system), 85, 89, 98
congruential logic, 68
Conjunction, 50
  Property of, 50
conjunctive logic, 50
consequence relation, 25
constructive falsity, logic of, 106
continuous (Leibniz operator), 60
continuous (Tarski operator), 66
contraposition rule, 94
correspondence theorem, 23, 60
Cut rule, 77
Czelakowski, J., 1, 2, 5, 6, 12–14, 36, 59, 60, 66, 68, 71, 86, 93, 102

Da Costa, N.C.A., 105
De Morgan abstract logics, 111
De Morgan algebra, 112
De Morgan lattice, 111
De Morgan Laws, 112
De Morgan monoid, 122
De Morgan semigroup, 122
Deduction Theorem, 51, 116
Deduction-Detachment Theorem, 51
deductive filter, 27
derivable sequent, 77
derived rule (of a Gentzen system), 76
detachment, 51
Disjunction Property, 53
Disjunction, Weak Property of, 110

disjunctive logic, 54
distributive abstract logic, 109
distributive lattice, 107
  pseudo-complemented, 113
Double Negation, Property of, 112
duality, 17
Dunn, J. M., 111

equation, 16
equational consequence, 60, 83
equational theory, 83
equivalent algebraic semantics (of a Gentzen system), 84
equivalent quasivariety semantics, 60
equivalential fragments, 106
equivalential logic, 36
extension, 18
  axiomatic, 18, 26
extension algebra, 118

Figallo, A., 125
filter
  deductive, 27
  implicative, 115
  Leibniz, 64
  logical, 27
  of a lattice, 108
  of a matrix, 16
Fine, K., 122
finer, 18
finitary, 17, 26
finitary part of an abstract logic, 80
formula, 15
  atomic, 15
four-valued logic of Belnap, 111
Frege, G., 68
Frege hierarchy, 68
Frege operator, 46
Frege relation, 46

Fregean logic, 68
    fully, 68
Fregean variety, 68
full model, 33
    basic, 34
fully adequate, 81
fully Fregean logic, 68
fully selfextensional logic, 48

𝔊-algebra, 80
generalized matrix, 18, 33
Gentzen calculus, 75
Gentzen system, 76
    algebraizable, 84
    algebraization, 75, 83
Gentzen-style rule, 45

Halmos, P.R., 14, 118
Harrop, R., 5
Herrmann, B., 25, 39, 62, 71
Heyting algebra, 116
hierarchy
    Frege, 68
    Leibniz, 29
Hilbert algebra, 99, 115
Hilbert-style rule, 46
homomorphism, 15

Identity Law, 123
implication algebra, 115
implicational logics, 105
implications-less fragment of IPC, 113
implicative filter, 115, 117
inconsistent element, 56
inconsistent set, 94
inductive, 17
interpretation, 16
introduction of modality, 57
intuitionistic logic, 116

equivalential fragment, 106
    positive fragment, 115
intuitionistic propositional logic
    implicative fragment, 115
Intuitionistic Reductio ad Absurdum,
    55
isomorphic abstract logics, 21
isomorphism theorem, 6, 8, 22, 41,
    107

K (modal logic)
    extensions of, 120
K-congruence, 15
Kleene algebra, 113
Kleene lattice, 112
Kleene's strong three-valued logic,
    112
Knuth, D.E., 14

Lamport, L., 14
lattice, 110
    De Morgan, 111
    distributive, 107
    Kleene, 112
    relatively pseudo-complemented,
    115
Leibniz, G.W., 16
Leibniz congruence, 16
Leibniz filter, 64
Leibniz hierarchy, 29
Leibniz operator, 17
Lemmon, E.J., 118
Lindenbaum, A., 29
Lindenbaum-Tarski algebra, 29
Lindenbaum-Tarski quotient, 29
logic, 9, 26
    1-equivalential, 71
    abstract, 17
    algebraizable, 60

almost inconsistent, 60
congruential, 68
conjunctive, 50
defined by a Gentzen system, 77
defined by an abstract logic, 31
disjunctive, 54
equivalential, 36
finitary, 26
four-valued, 111
Fregean, 68
fully Fregean, 68
fully selfextensional, 48
implicational, 105
many-valued, 105
modal, 57, 117
non-monotonic, 9
non-pathological, 59
of constructive falsity, 106
of lattices, 110
of paradox, 105
of positive implication, 99, 114
paraconsistent, 56, 105, 124
positive, 115
preserving degrees of truth, 105
protoalgebraic, 59
pseudo-axiomatic, 70
purely inferential version, 62
regularly algebraizable, 71
selfextensional, 48
sentential, 9, 25
sequential, 75
strongly algebraizable, 93
strongly selfextensional, 48
substructural, 9
three-valued, 112
two-valued, 68
weakly algebraizable, 66
with semi-negation, 106
logical congruence, 18

logical filter, 27
logical morphism, 20
logical quotient, 21
logical system, 26
Loureiro, I., 125
Łukasiewicz, J., 105

Makinson, D., 5
many-valued logic, 105
matrix, 16
for S, 27
generalized, 18, 33
reduced, 17
matrix congruence, 16
Meyer, R.K., 122
modal logic, 57, 105, 117
normal, 118
quasi-classical, 117
quasi-normal, 118
strong version, 118
tetravalent, 125
weak version, 118
with intuitionistic base, 120
modality, introduction of, 57
model, 32
basic full, 34
full, 33
of a Gentzen system, 78
Modus Ponens, 51
monadic Boolean algebra, 118
Monteiro, A., 6, 125

natural deduction, 5
Necessitation, Rule of, 118, 125
Nelson, D., 106
Németi, I., 66
non-modal reduct, 119
non-monotonic logics, 9
non-pathological logic, 59

normal modal algebra, 118
normal modal logic, 118

open problem, 14, 40, 48, 72, 75, 92, 101
open set, 119

paraconsistent logic, 56, 105, 124
    maximal, 123
paradox, logic of, 105
Peirce's Law, 115
Pigozzi, D., 1, 2, 5, 10, 12–14, 16, 25, 29, 59, 60, 66, 68, 71, 75, 83, 84, 86, 93, 94, 102
positive implication algebra, 99
positive implication, logic of, 114
positive logic, 115
Priest, G., 105
projective generation, 20
proof by cases, 53
Property of Conjunction, 50
Property of Disjunction, 54
protoalgebraic logic, 59
provability logic, 120
pseudo-axiomatic logic, 70
pseudo-Boolean algebra, 116
pseudo-complemented distributive lattice, 113
Pynko, A., 124

quasi-classical modal logic, 117
quasi-Hilbert algebra, 127
quasi-normal modal logic, 118
quotient matrix, 16

R-algebra, 122
Rasiowa, H., 1, 6, 36, 106, 114, 120
reduced abstract logic, 23
reduced matrix, 17
Reductio ad Absurdum, 55

reduction
    of a matrix, 17
    of an abstract logic, 23
reduction process, 23, 44, 49
regularly algebraizable logic, 71
relatively pseudo-complemented lattice, 115
Relevance Logic, 122
Relevance Principle, 122
Replacement (rule of $\models_K$), 83
residuated lattices
    logics of, 105
Routley, R., 122
rule
    derived, 76
    Gentzen-style, 45, 76
    Hilbert-style, 46
    structural, 76
Rule of Necessitation, 57, 118

S-algebra, 36
selfextensional, 48
selfextensional logic
    fully, 48
    strongly, 48
semantics, 30, 31
semi-negation, logic with, 106
semilattice, 90
sentential logic, 9, *see also* logic
sequent, 76
    derivable, 77
    of $\mathfrak{S}$, 77
sequential logic, 75
Sette, A.M., 123
Sette algebras, 124
S-filter, 27
Shoesmith, D.J., 5
S-matrix, 27
Smiley, T., 5, 19, 33

strong version of a modal logic, 118
strong version of an abstract logic,
   119
stronger, 18
strongly adequate, 81
strongly algebraizable logic, 93
strongly selfextensional logic, 48
structural rule, 76
substitution, 16
substitution invariance, 26
substructural logics, 9
Suszko, R., 5, 20, 68
Suszko's rules, 68
Symmetry (rule of $\models_\kappa$), 83

Tarski, A., 4, 10, 19, 29, 118
Tarski congruence, 18
Tarski operator, 19
Taylor, P., 14
tetravalent modal algebra, 125
tetravalent modal logic, 125
theorem, 17
theory, 17, 26
three-valued Łukasiewicz algebra, 125
topological Boolean algebra, 118
Transitivity (rule of $\models_\kappa$), 83
translation, 66, 84
trivial element, 56
(**t**,**s**)-equivalent, 84
turnstile, 76
two-valued logic, 68
type of a Gentzen system, 76, 77

variable, 15
Variable-Sharing Property, 122
Verdú, V., 6, 11, 14, 20, 37

Wajsberg, M., 105
Weak Contraposition, Property of, 112

Weak Property of Disjunction, 110
weak version of a modal logic, 118
Weakening rule, 76
weaker, 18
weakly algebraizable logic, 66
Wójcicki, R., 4, 5, 10, 33, 48, 122

Lightning Source UK Ltd.
Milton Keynes UK
UKOW02n1159080317

296147UK00007B/170/P